PRAISE FOR CHRISTOPHER DEWDNEY

"As you read these pages, your life will change, because the way you see half of it will change. The night we're all familiar with will emerge as a fresh thing, deeper, fuller, older, younger, more evocative, more intimate, larger, more spectacular and, yes, more magical, and much more thrilling."

— Margaret Atwood on *Acquainted with the Night*

"Tautly written in a highly condensed yet personable voice, this tour of the manifold nocturnal realm is a superbly meticulous feat."

— *Publishers Weekly* (starred) on *Acquainted with the Night*

"Keep this book by the bed for the lonely vigils; it will at least stop you envying those asleep — they're missing it all."

— *The Times* on *Acquainted with the Night*

"Dewdney's capturing of a zone of knowledge is a public event."

— *Books in Canada* on *Acquainted with the Night*

"Christopher Dewdney is a poet, one of the most remarkable working in Canada now."

— *Saturday Night* on *Acquainted with the Night*

"Nothing escapes Dewdney's oddly polymath intelligence."

— *Ottawa Citizen* on *Acquainted with the Night*

"*Soul of the World* demonstrates how poets can help us rediscover the wonders that lie hidden within the sciences."

— *Vancouver Sun* on *Soul of the World*

"A brilliant, brilliant book."

— Matt Galloway, *Here and Now*, CBC Radio, on *Soul of the World*

"He writes about time with the agility and insight of the poet he is, and with the clear exposition of an engaging and passionate teacher."

— *Globe and Mail* on *Soul of the World*

"These poems have a haunting, stately quality: they truly suggest the 'mystery/of everything' as manifested in the act of love."

— *Globe and Mail* on *Signal Fires*

"Dewdney has undergone a transformation; his poetry has taken on greater humanity and been touched by love, while still in touch with the gods."

— *Quill & Quire* on *Signal Fires*

"[His] poems have a painful, visceral quality to them that rasps against the cerebral and genteel aspects of their presentation. They are about the redemption of human misery magnified to the highest power."

— *Eye Magazine* on *Signal Fires*

"[He has] fashioned a body of work that is original, challenging, witty, stylistically versatile and remarkably cohesive."

— *Books in Canada* on *Signal Fires*

"[Dewdney's] insights are beautiful, provocative and haunting."

— *American Book Review* on *The Natural History*

MILES

THE EPIC DRAMA OF OUR ATMOSPHERE AND ITS WEATHER

CHRISTOPHER DEWDNEY

ECW

Published by ECW Press
665 Gerrard Street East
Toronto, Ontario, Canada, M4M 1Y2
416-694-3348 / info@ecwpress.com

Cover design: Michel Vrana
Author photo: © Greig Reekie
Editor: Susan Renouf

LIBRARY AND ARCHIVES CANADA
CATALOGUING IN PUBLICATION

Dewdney, Christopher, 1951–, author
18 miles : the epic drama of our atmosphere and its weather / Christopher Dewdney.

Includes bibliographical references and index.
Issued in print and electronic formats.
ISBN 978-1-77041-346-7 (softcover).
ALSO ISSUED AS: 978-1-77305-223-6 (EPUB),
978-1-77305-224-3 (PDF)

1. Weather—Miscellanea. 2. Weather—Popular works.
I. TITLE. II. TITLE: Eighteen miles.

QC981.2.D49 2018 551.6
C2018-902515-8 C2018-902516-6

The publication of *18 Miles* has been generously supported by the Canada Council for the Arts which last year invested $153 million to bring the arts to Canadians throughout the country, and by the Government of Canada. *Nous remercions le Conseil des arts du Canada de son soutien. L'an dernier, le Conseil a investi 153 millions de dollars pour mettre de l'art dans la vie des Canadiennes et des Canadiens de tout le pays. Ce livre est financé en partie par le gouvernement du Canada.* We also acknowledge the Ontario Arts Council (OAC), an agency of the Government of Ontario, and the contribution of the Government of Ontario through the Ontario Book Publishing Tax Credit and the Ontario Media Development Corporation.

ONTARIO ARTS COUNCIL
CONSEIL DES ARTS DE L'ONTARIO
an Ontario government agency
un organisme du gouvernement de l'Ontario

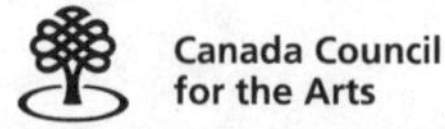

Conseil des Arts du Canada

PRINTED AND BOUND IN CANADA

PRINTING: MARQUIS 5 4 3 2 1

INTRODUCTION

"Sunshine is delicious, rain is refreshing, wind braces us up, snow is exhilarating; there really is no such thing as bad weather, only different kinds of good weather."

JOHN RUSKIN

You'd never guess it but we live in a world flatter than a sheet of paper. Shrink the Earth to the size of a basketball and our atmosphere would be as thick as a layer of food wrap. The oceans likewise. Two of the most critical elements for our survival, water and air, are relatively scarce commodities. We are like microorganisms living in an evanescent fluid film, a dampness that would burn off like morning dew if the sun increased its solar output by just 15 percent.

Astronauts know all of this. From the space station, they see the tops of clouds spread out at the surface of the atmosphere like smoke beneath glass. And slipping back under that thin blanket of air is a real challenge. Their reentry angle can be no shallower than 5.3° and no steeper than 7.7° — too shallow and they will ricochet into deep space, too steep and they will burn to a crisp.

Yet it's also a question of scale. The vantage from here, from Earth's surface, is of another order entirely. The sky seems to go on forever. When a waning moon shines by day, it looks to me as if it's suspended in the same blue atmosphere I breathe. No wonder Icarus dreamed of flying to the sun. And the immeasurable vastness of clouds, taller than mountains, what could contain that immensity?

For us, the atmosphere is a theater beyond reckoning, a massive, transparent stage for the drama of the skies. Every sunset is a light show; every storm a nail-biting, colossal thriller. Weather inspires our emotions and sometimes seems to reflect them. There is nothing more romantic than a rainy evening for newlyweds on their honeymoon, and how many philosophers have paced through windy streets deep in thought?

When I was a boy, the wind was a mood, a way of being, a kind of delirium that called me out of my house. I raced the leaves blowing along the street or stood at the edge of the ravine to hear the wind's soft thunder in my ears. Clouds were another mood. At sunset, they transformed into dreamlike landscapes inviting the secret empire of night. I was fascinated by weather. Every season was a new universe, the next chapter in an epic story I made up as I went along.

January found me at a research station in Antarctica in the ravine behind my parents' house. There I braved subzero blizzards to map glaciers with my special team of explorers, handpicked from neighborhood friends. Once, with amazing luck, we unearthed the frozen carcass of a mammoth, and on another expedition there was a time warp in the middle of a particularly dense snowfall, and we came face-to-face with a snarling saber-toothed tiger that charged out of the blizzard. Fortunately, we survived.

One hot July, there was a softball game followed by an expedition to explore the upper reaches of the Amazon in the same ravine — the raucous calls of howler monkeys echoed through the rainforest while treacherous Komodo dragons rustled in the undergrowth. We came upon the forest trails of lost tribes, sometimes catching glimpses of their ocher-painted skin as they disappeared around the bend of a trail. Of course, since then I've discovered that mammoths didn't visit Antarctica, nor do Komodo dragons inhabit Brazil, but I've never lost my deep connection with climate and weather.

When I was a little older, in my early teens, I was fascinated by weather forecasts. Forecasters were scientific magicians who could conjure storms out of a sunny afternoon. From *Weather: A Golden Nature Guide*, a book my parents bought me, I began to learn the weather signs: a ring around the sun meant rain within one to two days; earth glow on the moon (when the dark side of a half-moon is faintly visible) was the reflection of masses of white clouds to the west, an almost sure sign of rain to come. There were illustrations of hurricanes and tornadoes and sun dogs. Now I was really hooked.

Then one afternoon, while leafing through my new copy of the Edmund Scientific mail-order catalog, among the usual assortment of tempting items — ant farms, glow-in-the-dark stick-on stars, aquariums, test tubes — I noticed a new item, a complete home weather station that included an outdoor anemometer (those whirling wind cups that measure wind speed). I had to have it. I saved my allowance and made extra money doing yard work.

When the package arrived, it was a little smaller than I expected, but everything was in there, including the glorious anemometer with its three wind cups. I blew on them and they twirled obediently, orbiting the little mast. There was also a separate spar for the wind vane, and both had connections for wires that led to my indoor instrument panel. All I had to do was install the wind vane and anemometer high enough to give accurate readings, which wouldn't be easy. It meant I had to make a trek to the top of my parents' house.

The afternoon of the installation was cool and windy. I climbed out of an attic window that was barely big enough to squeeze through and then screwed the base for both masts into the wooden gable above the dormer at the peak of the house. I imagined being captured by a *National Geographic* photographer as I braved the harsh mountain gales to install my weather station. After setting up the anemometer and wind vane, I connected the wires and threw the loose coils over the eaves in the general direction of my bedroom window below. Then I clambered back inside.

In my bedroom, I snared the dangling wires with a rake and pulled them through the window. I'd already installed the two wall-mounted, battery-powered weather gauges that the wires would connect to, one for wind velocity and the other for wind direction. Then came the moment of truth — if nothing happened, if the instrument panel was dead, I'd have to trek back to the roof to check the connections. I hooked up the wires, and the gauges danced to life.

The directional gauge was a circular compass with a little arrow that pointed out the wind direction in tandem with the vane on the roof mast. The velocity gauge was horizontal with a needle indicator — like an old-fashioned speedometer in an automobile dashboard — that showed a wavering wind speed of about 20 miles per hour. I was euphoric. Along with the barometer and window thermometer that I'd previously mounted, I now had a professional indoor weather station. I could take instrument readings from the comfort of my own bedroom no matter what the weather outside, and, more importantly, I could make my own weather forecasts, going mano a mano against the evening news' weatherman.

By combining weather signs with instrument readings, I became a pretty good forecaster. I learned that a halo around the moon at night along with falling barometric pressure meant that it would probably rain within 18 to 48 hours. When an east wind shifted to the west and the cloud bases got higher and the barometer was rising, fair weather usually followed. In the winter, a north wind that shifted counterclockwise to become a west wind and then a southerly wind meant that snow was likely within a day.

Later I discovered that I could make pretty good predictions — especially of stormy weather — using only wind direction and my barometer. If the wind was blowing out of the south and then shifted to the east and my barometer was 29.8 inches or below and falling rapidly, a severe storm was imminent. The same was true if the wind shifted from east to north, especially in the winter and my barometer again was showing 29.8 inches or below and falling rapidly. I usually compared my results to the evening news' weather report. I wasn't always right — there were some things I just couldn't see coming without a satellite view and upper-atmosphere readings — but I did pretty well considering.

But for all the quantitative data I was now receiving, my love of weather remained visceral, aesthetic even. The instrument panel just underlined the meteorological drama. A howling gale, even if gusts were measured at 50 miles per hour, was still a howling gale, with all the excitement of the wind roaring through the trees and garbage cans blowing down the street. Somehow the science permitted an illusion if not of control then at least perhaps a complicity of sorts. I was part of the weather.

Today I like to think of myself as a connoisseur of weather, an epicurean of hourly changes. Perhaps it's a consequence of being a writer, or maybe I'm just meteorologically sensitive, but I'm very susceptible to the moods of weather. I revel in certain hot, overcast August afternoons with a ceiling of featureless, rainless stratus clouds. It's the brightest light possible without casting any shadows, only a gathering of darkness under the parked cars or trees in the park. I like a similar sky on October afternoons when the undersides of the clouds are quilted, and the gray light seems to amplify the fiery reds and oranges in the autumn foliage.

There's magic to urban evenings just after the sun sets and the city lights the bottoms of scattered cumulus clouds. They become islands between which stars ride an indigo blue ocean. And in July there are windy, hot summer afternoons, clear and dry, sometimes followed by equally windy summer nights where even the Milky Way seems to be adrift. I have seen sunsets as astonishing as fireworks, like surreal Sistine ceilings that stretched from horizon to horizon, and I remember foggy mornings like mysteries that dissolve the world. As T.S. Eliot wrote about ocean fog in his poem "Marina," "What seas what shores what gray rocks and what islands / What water lapping the bow / And scent of pine and the woodthrush singing through the fog."

What exquisite atmospheric nuance — a boat in the fog, where scent and song are the only beacons. Eliot's fog conceals our highest spiritual aspirations and yet also evokes our devastating ignorance. As a species, we have so much left to understand and yet our yearning is our beacon. In a way, we're like astronauts riding flaming ships through the sky on their return to Earth; we can only have faith. The astronauts know our atmosphere is a narrow, fragile margin, but they also know that it's a magnificent realm — at once gorgeous, terrifying, capricious and elusive.

1
STORMY WITH A CHANCE OF LIFE
THE IMPROBABLE BIRTH OF OUR ATMOSPHERE

The fortuitous creation of Earth took place about 4.5 billion years ago. A billion years is a long while. To get an idea of the immensity of such a span, imagine time could be condensed into a substance and that each year deposited a gram — the weight of a ballpoint pen cap or a dollar bill. If you added each year to the next, a decade would weigh 10 grams and a normal human life span would be 80 to 90 grams, about the weight of a chocolate bar.

Suppose we kept going — every year backward adding another gram to the lump — then 2,000 years, back to the time of the

Roman Empire, it would weigh an easily hefted two kilograms, or as much as a small sack of potatoes. Spool back 200,000 years to the first anatomically modern humans, and you're getting close to the limit of what an Olympic weightlifter can clean and jerk, around 200 kilograms.

Go further back to the dinosaurs, 60 million years ago, and your gram-a-year interest account cashes out at 53 tons, about the weight of a small diesel locomotive. Five hundred million years ago, at the beginning of what paleontologists call the Ordovician period, when our oceans were populated with trilobites and crinoids, the yearly deposits would weigh in at about 50,000 tons, or the approximate weight of three Ohio-class nuclear submarines. Four and a half billion years ago, when our planet first coalesced from primordial dust, your gram-a-year investment would weigh as much as an asteroid, one big enough to wipe out life on an entire continent.

Back then, our planet collided with impactors the size of our time-deposit asteroid every few hundred million years or so. Despite this constant pummeling, our molten planet had enough time to let gravity sort its constituent parts into layers — iron at the core and lighter minerals and elements arrayed above. The lightest of these, the gases, formed the top layer. At the time, these were hydrogen and helium, and they formed Earth's first atmosphere. Think of the Hindenburg disaster: just one match and the whole works would've exploded. But you could strike a thousand matches 4.5 billion years ago without producing a single spark. There was no oxygen. Anyway, why would you bother? You'd be asphyxiating. And with all that helium around your squeaky last words would be comically high pitched, in a macabre sort of way.

But helium is a fickle gas, and it didn't stick around long. Less than a hundred million years after Earth's formation, most of it had escaped Earth's gravity and fled into space. Unbonded gaseous hydrogen followed helium shortly afterward, leaving behind an atmosphere that had transformed into a pungent mixture of nitrogen, water vapor, carbon dioxide and hydrogen sulfide. Eau de rotten eggs. Beneath this odiferous miasma was a watery planet studded with a few transient islands of rock. It took another 300 million years for the Earth's crust to stabilize into a thin layer of congealed lava over the primeval magma, yet even then whatever proto-continents had managed to poke their landmass above the oceans were pelted by meteors and asteroids. In fact, every few hundred million years, when a particularly large asteroid struck, the oceans evaporated in the subsequent planetary inferno. For thousands of years afterward, the seas bided their time as atmospheric steam and only reformed when the surface of the planet had cooled to the point where rain no longer vaporized instantly on the red-hot surface but began to accumulate in puddles, lakes and finally oceans.

In the midst of these hydrogen tempests, meteor bombardments and constant volcanic eruptions, the most extraordinary development on Earth took place — self-reproducing organisms with rudimentary DNA appeared. And, as it turned out, these diminutive creatures packed quite an atmospheric punch.

THE PRIMORDIAL SOUP

Is there a hyperbole or superlative that can begin to capture how unlikely was the appearance of life? I think not. Life's emergence rivals, perhaps even surpasses, the sudden materialization of the universe itself, conjured *ex nihilo* over 10 billion years ago. But what is life? How can we characterize this special case of matter taking on such extraordinary abilities? Maybe I make too much of it. Perhaps life is, as a character drolly referred to it in Thomas Mann's *The Magic Mountain*, merely "an infectious disease of matter." But how did it start? How did inanimate molecules begin to copy themselves and persevere?

That's a question we have no detailed answer to. But we have a very informed general idea, and it goes something like this — life came from bad weather. It didn't begin on a calm planet with tranquil seas and light breezes; it started on a planet that had terrific storms, howling winds and waves hundreds of feet tall. Self-assembling molecules began to arrange themselves in the midst of volcanic eruptions and meteorite bombardments. They flourished in an agitated saline broth constantly electrified by lightning, scalded with lava and cooked by a young sun wielding dangerously high UV levels.

Alexander Oparin was the first scientist to envisage this alchemical, Frankensteinian jump-start to life. In his 1924 "primordial soup" theory, he speculated that ultraviolet light acting on elemental gases, liquids and solids in an oxygen-free environment created organic proteins, the basic constituents of life. Almost 30 years later, in 1953, Oparin's theory was vindicated by an ingenious and now famous laboratory experiment by Nobel laureate Harold Urey and his graduate student Stanley Miller. At the University of Chicago, they filled a series of beakers, glass tubing and electrical circuits with hydrogen, water and methane, seeking to reproduce the conditions on our planet as they were four billion years ago. For days, they zapped their broth with electricity to simulate the storms that raged across the ancient seas, and after only *a week*, 15 percent of the carbon in this shocked concoction had formed no less than 23 amino acids, the building blocks of complex life. They had proven that organic molecules could indeed have formed spontaneously from inorganic constituents.

There have been many critics of the theory since then, particularly creationists, of which one was my family's plumber, Gordon Lane. I remember watching him melt solder with a blowtorch to join two copper pipes under our bathroom sink when I was eight years old. Often he would stay for supper. He was a Jehovah's Witness with a Mensa IQ and, like my father, had a penchant for puns. He loved nothing better than to take on our agnostic family in fundamental arguments. He was particularly dismissive of the primordial soup explanation for the beginning of life. He knew the odds against assembling a simple cell were catastrophically immense, and he was right. A simple protein like collagen, for example, is a molecule with 1,055 sequences that have to be in exactly the right order to function. And that's just one of several hundred thousand proteins. He used a wonderful metaphor to underscore his argument against random mutations creating life-forms. "If I stood outside an auto wrecker's yard and threw rocks into the yard over the fence," he used to say, "I could stand there and throw rocks for a million years, and I'd never hear the sound of a car starting up on the other side of the fence."

Yet despite its critics, Oparin's theory still stands, mainly because there are so many naturally occurring amino acid proteins. It would appear that given enough time, in this case hundreds of millions of years, proteins could indeed have combined and gradually become more complex. Recently, the Miller-Urey experiment has been successfully revisited. Other researchers have added volcanic gases to the Miller-Urey mixture and they too have brewed amino acids as a result. Not only that, it appears that we have been importing some of our complex proteins, including amino acids, from outer space. A large meteorite that fell in Murchison, Australia, in 1969 was found to contain 20 types of amino acid with no terrestrial source. So if you add the infall of amino acids from meteorites and comets to the stew of proteins already brewing in the early oceans, then you have quite a broth of life-builders in the primordial soup.

But self-replicating proteins had to be reinvented many times over millions of years before one of them, again by chance, developed a membrane that gave it protection from the elements. It was the extraordinary good fortune of these proteins to emerge on a watery planet that orbited the sun at just the right distance — the Goldilocks Zone, as atmospheric scientists refer to it. Too close to the sun, like Venus, and water boils away; too far from the sun, like Mars, water freezes. And water, as it turns out, has a particular quality that jump-started intracellular transport and cell membranes.

Water is bipolar, not in the manic depressive sense, but in the electrical sense. One side of a water molecule has a positive charge, the other a negative charge. Like little magnets, they attract each other with just enough strength to stay grouped together but not so much as to turn into a solid. This makes water an excellent transport medium for dissolved minerals and chemicals. It's also why it has a meniscus, that layer of surface tension at the top of water you can see in an aquarium. Water molecules attract each other in all directions in deeper water, but at the surface they can only be attracted across the surface and downward, which aligns them into a temporary membrane. The first self-encapsulated proteins mimicked this property. Their membranes had water-loving molecules on the outside and water-repelling molecules on the inside. These joined in a circle to form a membrane that protected the delicate, nanomachinery of their interior.

Self-encapsulated proteins flourished and grew more complex, eventually crossing the line by which we define life a little more than four billion years ago, less than 500 million years after the birth of the planet. These first simple, single-cell creatures were called prokaryotes and used sulfates as a source of energy. They were anaerobic, meaning they flourished in the absence of free oxygen. Prokaryotes dominated the oceans for hundreds of millions of years. During their reign, unicellular life established itself and achieved planetary distribution, though a time traveler standing on the shore of that ancient ocean would see no evidence of life. Only a microscope would reveal the ubiquity of unicellular organisms. Anyway, you wouldn't have much time to collect samples because three billion years ago, when prokaryotic life had become firmly established, the environment was anything but temperate.

A TYPICAL WEATHER REPORT THREE BILLION YEARS AGO

First of all, the days were shorter. The Earth was spinning three times faster than it is now. A full day-night cycle was eight hours long, with a little more than four hours of darkness and four hours of pale sunlight because, even though UV levels were high, the young sun was fainter than today. You'd definitely have needed an oxygen mask — the atmosphere was almost entirely composed of carbon dioxide. And when the moon rose, you'd have known it. It was much closer to Earth and would have appeared 12 times larger than it does now. Today, the moon looks to be the same size as a dime held at arm's length. Three billion years ago, it would have looked the size of a cantaloupe. And you could hardly have called it moonrise. It would leap above the horizon and careen into the heavens, wheeling dizzily through the sky. Shadows cast by the moon, or the sun for that matter, would stretch and slide visibly as you watched them, like a time-lapse film.

Certainly, moonrise over the primeval ocean would have been a wondrous sight, but you wouldn't want to have been anywhere close to the water. In fact, the only safe vantage on the ocean would have been from the summit of a mountain somewhat inland. The tides were 1,000 feet high and arrived as quickly as a tsunami. Those prokaryotes living in the primeval oceans mustn't have had much rest.

Evolution was a slow-acting force at this time, but after many hundreds of millions of years, a momentous change finally did occur, a chance mutation that led to a new single-celled life-form, one that had a terrific edge over its anaerobic predecessors. Cyanobacteria. This newcomer took advantage of the relative abundance of carbon dioxide in the atmosphere as well as the sunshine. Cyanobacteria combined carbon dioxide and sunlight with water to produce carbohydrates for food. In essence, they survived exactly as plants do today. They were green, and, like plants, their unique metabolic process had a simple waste product — oxygen. Free oxygen, that formerly minor player in Earth's oceans and atmosphere, became a major one somewhere between 2.8 and 2.5 billion years ago.

TINY TERRAFORMERS

If humans eventually colonize other planets, they will rely on giant factories to process alien atmospheres into something breathable in a process called terraforming. Plans are already being drawn for the eventual terraforming of Mars. These mega-engineering projects will easily surpass any earthly achievements — the Pyramids, the Panama Canal, the Great Wall of China — but we have yet to build them.

Fortunately for us, Earth has already been terraformed. But huge machines didn't process our atmosphere; cyanobacteria did. Slightly less than three billion years ago, the dominant type of cyanobacteria lived in coral-like colonies called stromatolites. They formed knobby reefs in the oceans where they quietly bubbled away, releasing oxygen into the water. If you could stroll along the beach of a primeval ocean, you'd likely come upon wide, submerged ledges of these low reefs sitting just offshore, stretching alongside the ancient seaside as far as the eye could see. The air would be warm, but you'd still need an oxygen mask. The stromatolites and their allies had to pump out oxygen for hundreds of millions of years before the overflow leaked into the atmosphere.

Surprisingly, stromatolites have survived. They are the kings of living fossils. Nothing — not the tuatara lizard of New Zealand, unchanged for 100 million years, or even the coelacanth, the missing-link fish from Madagascar that looks today as it did 350 million years ago, or the sponges, with their billion-year heritage — holds a candle to the stromatolites, unchanged for 2.8 billion years. There is a thriving colony of them in Shark Bay on the west coast of Australia and another at Exuma Cays in the Bahamas. These unprepossessing, gray, blob-like rocks smeared with a thin layer of cyanobacterial cells — so small that one billion are contained in a square foot — were the dominant life-form on our planet for almost two billion years. It was almost as if life had stopped evolving.

THE OXYGEN CATASTROPHE

But while cyanobacteria were transforming the oceans, they were also killing off their predecessors. Oxygen was lethal for the pioneers of life on Earth — the prokaryotes and extremophiles that had flourished for a billion years. In a mere 300 million years, cyanobacteria had saturated the oceans with so much oxygen that 99 percent of the prokaryotic organisms died out in one of the greatest extinction events the Earth has known. Only a few survived at the bottom of oceans near thermal vents or buried under mud where oxygen could not find them. As a result, the Great Oxygenation Event is also called the Great Oxygenation Extinction Event or, more briefly, the Oxygen Catastrophe. Even so, the prokaryotes had a good run, dominating the planet for almost 400 million years. Their descendants still subsist today, buried deeply in rock or mud.

But those little bubbles of oxygen that cyanobacteria released not only exterminated most of the prokaryotes, they also initiated a vast, irreversible geochemical reaction. Underwater iron deposits began to rust for the first time in Earth's history. The oceans must have been tinged orange for millions of years during this great undersea rust bloom. The oxidizing iron left telltale bands in sedimentary rock that was laid down at the bottom of these newly oxygenated oceans. Today, geologists commonly find three-billion-year-old rocks with red band formations that contain layers of rust-dyed sediment, evidence of the first free oxygen on our planet.

For 300 million years, the cyanobacteria's steady output of oxygen was absorbed by iron and buried in ocean sediments. When all the available iron had bonded with oxygen, around 2.5 billion years ago, there was nowhere for the excess oxygen to go, so it bubbled out of the water and into the atmosphere. Oxygen levels in the atmosphere began to rise precipitously, triggering another oxidation event — all the exposed iron on dry land started to rust. Just like the banded marine sediments in ancient seafloor strata, these land-based layers can be plainly seen in rocks dating from this period.

From the perspective of an adjacent planet like Mars, the Earth would have undergone a Technicolor transformation. Within centuries of the first atmospheric oxygen, the continents would have changed color from brown and gray to bright terra cotta. By then, the oceans would have regained their original color, and Earth would have been a blue-and-orange planet twinkling in the Martian sky. This change of planetary color was testament to life's power. The primeval atmosphere had been transformed by life, and Earth's fate was now unique. It had diverged from a standard, geological planetary process and was now setting out on its own individual path. Life had begun to shape the face of the planet.

If our imaginary time traveler stood on the rust-red shore of that ancient ocean 2.5 billion years ago, she would no longer have needed an oxygen mask. The air would have been as deeply and fragrantly breathable as any seaside air today. Oxygen was in great supply, but there was not yet any multicellular life around to take advantage of its abundance.

There was a wrinkle though — that first whiff of oxygen was a cool one. Just as the continents began to turn rust red, ice caps appeared at both poles. Within a few thousand years, these polar ice caps accumulated into continental ice sheets that drove both south and north until only a very narrow band of ice-free ocean and land girdled the Earth at the equator. The first planetary deep freeze, the Huronian glaciation, was taking center stage. It lasted 300 million years and ended during a period of intense volcanic activity. Critically for us, the stromatolites and other cyanobacteria had survived in their equatorial refugia. But the Huronian glaciation, as it turns out, was just a warning shot over the bow of life's fragile boat: worse glaciations were to come.

As our imaginary time traveler standing on that shore 2.5 billion years ago could tell you, the atmosphere was similar to the atmosphere we have now, although carbon dioxide levels were much higher. Today the atmosphere is composed of 13 gases, of which two dominate — oxygen at 21 percent and inert nitrogen at 78 percent. Those ratios are important. Take oxygen, for instance. Every single percentage point over 21 percent increases the likelihood of forest fires by 70 percent. If oxygen ever reached 25 percent, all land vegetation — from the high Arctic to the equatorial rainforests — would eventually burst into flame in a raging, planetary wildfire. Nitrogen also sits at a sweet spot. If nitrogen levels fell to 75 percent, the climate would spiral into a deep freeze from which the Earth would never recover.

The other important gases are trace gases like argon at 0.9 percent and carbon dioxide at just 0.04 percent, neon at 0.001818 percent, hydrogen at 0.000055 percent, methane close to 0.00018 percent and helium at only 0.000524 percent. The other gases are very minor players, except for ozone, which, like carbon dioxide, has a disproportionate influence on Earth's habitability. Ozone forms a diaphanous, ethereal umbrella over the Earth, shielding the planet from damaging ultraviolet radiation. When scientific research proved that chlorofluorocarbon from spray cans and refrigerators was destroying the ozone layer, international legislation was enacted in just over a decade. It was an open-and-shut case. Without ozone, all earthly vegetation and most creatures would burn and mutate in the intense ultraviolet radiation. Despite that, ozone, along with radon, krypton, xenon and nitrous oxide, adds up to merely 0.000004 percent of the atmosphere.

Nitrogen is a majority shareholder in our atmosphere and yet, other than maintaining the Earth's moderate temperature, it's a silent partner. Certainly wine aficionados use pressurized containers of nitrogen to cap their opened bottles (apparently it works better than a vacuum seal), and filling car tires with nitrogen instead of air is a recent automotive trend, but compared with oxygen, nitrogen seems almost menial, commonplace. But don't be fooled; nitrogen has a cosmic pedigree.

If other planets harbor life, their atmospheres likely contain a significant percentage of nitrogen. It is the seventh most plentiful element in our universe and has been incorporated into every living thing on Earth. Its absence would doom us. Nitrogen makes up about 3 to 4 percent of the dry weight of all life and provides an essential element for cellular construction and amino acids. It is present as nitrates in animal waste and urine. Potash is essentially nitrogen, and we mine it because much of our food comes from plants that require nitrogen fertilizer.

But there really shouldn't be as much nitrogen as there is. Normally, nitrogen and oxygen react with each other, and, over the eons, they should have combined. Most nitrogen should be sequestered in the oceans as stable nitrate ions. Yet that hasn't happened. It's part of the counterintuitive miracle of our atmosphere. Something, and it's most likely life, is keeping the mixture from reacting.

Nevertheless, in terms of climate control, nitrogen, for all its ubiquity and abundance, is almost completely overshadowed by carbon dioxide in terms of punching above its weight. Carbon dioxide represents only 0.04 percent of our atmosphere, or approximately 400 parts per million (ppm). If you added the same proportion of the poison strychnine to water, you could drink gallons without the slightest effect. These concentrations have remained fairly stable over the past 400,000 years, ranging from 180 ppm during the height of glacial ages to 290 ppm during interglacial periods. Yet despite that scarcity, it is critical for regulating the overall surface temperature of the planet and is therefore completely essential for life on Earth.

Carbon dioxide is crucial for plant survival and therefore for all life-forms right up the food chain. After the first plants learned how to extract carbon dioxide from the atmosphere with their exquisitely complex nanomachinery, they enlisted photosynthesis to convert carbon dioxide into energy. Then they bootstrapped us out of the cradle of the ocean. Plants are our heroes. Today the global sum total of energy captured by photosynthesis is about 130 terawatts, six times more than the total amount of energy used by all human civilizations currently in existence. The carbon that's not used for energy builds the branches, roots, leaves, flowers and stems. This fixes the carbon, and when the plant dies, the carbon is sequestered in the soil, eventually compressed into rock, which is then, over millions of years, subducted into the molten interior of our planet to be later released by volcanoes. A cycle of air, life, rock and fire.

The world's volcanoes, on average, release about 130 to 230 megatonnes of carbon dioxide yearly. It sounds like a lot, but it really isn't next to the staggering amount that photosynthetic organisms contribute. Decaying vegetation, both underwater and on land, creates carbon dioxide. Seaweed and plankton in the oceans produce about 332 gigatonnes of carbon dioxide yearly, yet even that pales in comparison to land vegetation, which produces a staggering 439 gigatonnes of carbon dioxide every year. By contrast, humans only pump 29 gigatonnes into the atmosphere yearly. Problem is, that small percentage we're adding doesn't have a natural carbon sink to neutralize it. The 771 gigatonnes from the oceans and land, as well as the 180 megatonnes (on average) from volcanoes, are all accounted for within our planetary carbon cycle. So our surplus is accumulating. We're playing with fire, literally, by burning sequestered carbon and adding it to the atmosphere. In the mid-twentieth century, carbon dioxide levels were at 320 parts per million (ppm); now they are climbing over 400. We've been doing this for quite some time.

And for all that, in geological terms, our dubious effort to add more carbon dioxide into the atmosphere is doomed. Carbon dioxide concentrations have been steadily decreasing since their highest levels in the primordial atmosphere, when it was *the* dominant gas. Gas newbies, oxygen and nitrogen, pushed out carbon dioxide to the extent that by the Cambrian period, 500 million years ago, carbon dioxide was already a trace gas, with concentrations at about 7,000 ppm. Concentrations of carbon dioxide decreased to 3,000 ppm during the Jurassic and Cretaceous periods, more than 60 million years ago. Then they fell lower, 34 million years ago, to 760 ppm. You can see where this is going. Today carbon dioxide concentrations stand at approximately 400 ppm. Over the long term, in a hundred million years or so, one of the most essential gases for the continued existence of life is going to run out. But that doesn't let us off the hook, not by a long shot. Here and now, carbon dioxide management is a global dilemma.

There's one more atmospheric player I haven't mentioned yet, mainly because it isn't a gas. Water. It occupies about 2 percent of the atmosphere in the form of water vapor and clouds, though that 2 percent is distributed very unevenly. Warm air holds more moisture than cold air, so there's more atmospheric water aloft over the tropics than above the polar regions. In its global entirety, on average, the atmosphere holds about 375,000 trillion gallons of water. There is an ocean above our heads.

2

THE WILD BLUE YONDER

THE LAYERS OF THE ATMOSPHERE

When I was a child, I would sometimes climb up onto my parents' garage roof on late summer afternoons and lie down on the warm asphalt shingles. There was a feeling I was after, a sort of giddy, floating sensation when I stared straight up. I'd imagine the garage was gone and that I was alone, suspended in the endless blue sky that surrounded me. No neighborhood, no city, no planet Earth, just blue, blue atmosphere and the sun. Well, there were birds winging by me in my azure world, but they were fellow travelers.

I would sometimes invite my best friend up to share my fantasy, though he would get vertigo if he got too deeply into the vision. In fact, it was the vertiginous sense of endlessly falling upward into an infinite, warm universe of blue sky that I was secretly addicted to. The sky was a sea of warm air in which I hovered. And so it is. We live at the bottom of an ocean of air, 5,200 million million tons to be exact, or 25 million tons piled on every square mile of the planet. If the atmosphere were compressed into a slab of granite, it would be 2,000 miles long, 1,000 miles wide and half a mile thick. It sounds like a lot but Earth's atmosphere is pasted in an alarmingly thin layer — 99 percent of it lies within 18 miles of the surface. Earth's transparent skin. Without it we couldn't exist.

Earth is not unique in the solar system in terms of having an atmosphere. All the planets have atmospheres except Mercury, which has the misfortune of being too close to the sun. A few of the 173 moons that orbit other planets in our solar system also have atmospheres, but most moons, including our own, do not. Not a whiff, not a puff, not a wisp of gas. The moon's airless surface is completely exposed to the extreme temperature swings of outer space. At lunar noon, the thermometer often reaches 123°C, so if you were to set a glass of water on the regolith (the lunar soil), it would boil away in a few minutes. At night, the temperature plunges to -181°C. If you were an astronaut there in the starry darkness and if you had the suicidal inclination to remove your helmet, the carbon dioxide in your last exhalations would freeze and fall as snowflakes of dry ice.

We're lucky here on Earth not only to have an atmosphere but a relatively benign one. In fact, we can get down on our knees and thank our atmosphere for protecting us — and not just from such temperature extremes. It also intercepts about 99 percent of meteorites and a lot of downright hostile subatomic particles. And if that isn't enough, Earth's atmosphere has just the right balance of oxygen and nitrogen for us to breathe comfortably.

But it *is* a thin layer At 19 miles high, the windows of ascending spacecraft darken and stars begin to appear. The unofficial border of outer space, or at least the airless vacuum part of it, is only the distance of a big-city crosstown excursion, albeit a vertical excursion that consumes an Olympic swimming pool's worth of rocket fuel rather than a dollar of gasoline.

In horizontal terms, we could walk the distance in about six hours. Climbing would be a whole other matter. The peak of Mount Everest, at 5 1/2 miles (29,029 feet), only gets about a third of the way. The lucky few that summit Everest, exhausted and light-headed, must get a dizzy sense of what space feels like. At the mountain's peak, the sky is a dark sapphire blue and there's barely enough oxygen for a human to survive. It's not because oxygen levels drop disproportionately compared to the rest of the atmospheric gases up there — oxygen remains at 21 percent — it's because the air itself is thinning out. With less atmospheric pressure, the lungs cannot extract the same amount of oxygen with each breath.

Airliners normally fly one and a half miles above the height of Everest, cruising at 35,000 to 40,000 feet, or 7 miles high. They're a little closer to the edge of the atmosphere, which is why passengers breathe a pressurized atmosphere. It might also be why a few cocktails imbibed en route suddenly gang up on unwary drinkers when the plane lands and the cabin is flushed with normal atmospheric pressure.

The early aeronauts of the nineteenth century didn't have in-flight service during their daring ascents. And they certainly had no onboard oxygen or even a closed cabin when they piloted their balloons into the upper atmosphere. They hadn't a clue what they were going to encounter up there — they ascended literally into the unknown. Scientists at the time didn't how high the atmosphere extended or if its composition changed with altitude. From mountaineering accounts, early aeronauts knew that temperature dropped the higher you went and that oxygen levels also fell, but that was all. There was nothing for it but to send someone up, and the only way was to hang from a fragile balloon filled with coal gas.

UP, UP AND AWAY

The grand exploration of the atmosphere began in 1783 when Jean-François Pilâtre de Rozier and François Laurent, marquis d'Arlandes, first rose above France in a hot air balloon. The ballooning craze, for those who could afford it, spread quickly throughout Europe, and by the beginning of the nineteenth century, ballooning had become a science, and the race to probe the atmosphere was on.

The first scientist to go aloft was the great French chemist Joseph-Louis Gay-Lussac, whose specialty was gases, particularly how the volume of a gas changes with temperature and, as he would soon find out, with height. It was during his second flight in a hydrogen balloon in September 1804 that he set an altitude record, reaching a height of 23,018 feet (4.3 miles). Gay-Lussac carried a barometer, thermometer, hydroscope, compass and flasks to capture samples of air. From air samples taken during the ascent, he discovered that the composition of the high-altitude atmosphere was the same as that at the surface; it was just thinner. One of his most useful discoveries was the confirmation that the temperature dropped by almost 1°C for every 328 feet in elevation. He was also the first human to experience the beginnings of decompression — suffering a tremendous headache at the peak of his ascent. It would be more than 50 years before Gay-Lussac's altitude record was broken by English aeronauts James Glaisher and Henry Coxwell on the afternoon of August 18, 1862.

James Glaisher was a self-taught mathematician and meteorologist who had practically grown up in the Royal Observatory at Greenwich Park. Years later, when the appropriately named Sir George Biddell Airy was appointed Astronomer Royal in 1835, one of his first acts was to make Glaisher superintendent of the new magnetical and meteorological department. Glaisher was an excellent scientist and had a knack for reducing almost any phenomenon into numbers. (He once devised a formula for what he insisted was the perfect cup of tea.) But above all, he was a dedicated meteorologist.

He was instrumental in founding the British Meteorological Society in 1850 and then on August 8, 1851, as part of the Crystal Palace of Industry exhibition, he published the very first British daily weather map from telegraphically collected data. It was a breathtaking miracle of scientific technology, and the public was enthralled.

Glaisher was fascinated by the dew point, the temperature at which condensation forms out of atmospheric water vapor and he hoped, with this voyage into the upper atmosphere, to find out how the dew point might change with altitude.

The pilot of their vessel, Henry Coxwell, was an experienced balloonist, a veteran of vertical ascents. They were an optimal crew for such a mission. On a hot August day, they clambered aboard and sat amid Glaisher's scientific equipment in a basket (which balloonists of the time referred to as a car) suspended below the largest balloon that had ever been inflated. It was called the *Mammoth*, and it was filled to bursting with 90,000 cubic feet of coal gas. Everything they needed was stored aboard, including a first-aid compartment containing only a pint of brandy.

Once the balloon was straining at the tethers, all that Coxwell and Glaisher had to do was cut them. The balloon rose quickly, and in a little more than 20 minutes, it had risen three miles above the Earth's surface. To get even more elevation, they threw some ballast overboard. (I wonder what happened to those sandbags. Dropped from that elevation, their terminal velocity would be around 330 miles per hour, yet there seem to be no contemporary accounts of holes smashed through the roofs of vicarages or complaints about flattened livestock.)

During their ascent, Glaisher took readings from his instruments, including a barometer that also acted as an altimeter. About

30 inches of mercury is normal for sea-level air pressure. Previous balloon ascents and experiments that involved transporting barometers up mountains had proven that air pressure drops two inches for every half mile of elevation. That gave our Victorian balloonists a fairly accurate idea of how high they'd ascended. By the time the barometer had dropped to 11.5 inches, or about five miles high (26,400 feet), the temperature inside their car was -23°C, and the sky had turned a dark, Prussian blue.

At this point, Glaisher began to pass out, losing the ability to move one of his arms. He later wrote in his diary, "I then tried to move the other arm, but found it powerless also. I next tried to shake myself, and succeeded in shaking my body. I seemed to have no legs. I could only shake my body. I then looked at the barometer, and whilst I was doing so my head fell upon my left shoulder . . . and then I fell backwards, my back resting against the side of the car, and my head on its edge. In that position, my eyes were directed to Mr. Coxwell in the ring." And what he saw momentarily before he completely lost consciousness panicked him.

George Coxwell, realizing they'd gone too high, was also suffering from extreme oxygen deprivation, but he would prove his balloonist mettle. While Glaisher was passing out, Coxwell had clambered out of the car and up into the lattice of ropes suspending the car from the balloon. The valve that let the air out of the balloon (so they could descend) had been twisted by one of the guy ropes and was out of reach. Coxwell managed to snare it and pull it back into the car, but by then he'd also lost the use of his arms. Holding the valve with his teeth, he gave it three tugs and the gas began to escape. Then he collapsed on the floor of the car as they started their descent. In a few minutes, Glaisher regained consciousness and resumed his observations until they landed.

Glaisher wrote, "When we dropped it was in a country where no accommodation of any kind could be obtained (Shropshire), so that we had to walk between seven and eight miles." Not exactly a hero's welcome. Although they didn't know it at the time, they had sailed up through the entire troposphere to the edge of the stratosphere.

The troposphere, about seven miles thick, is where all weather happens: clouds, jet streams, rain and hurricanes. The stratosphere is a much more tenuous layer of atmosphere. Its lowest regions, right where it meets the troposphere, are very cold, around -60°C. Glaisher and Coxwell could have told you that. It's why when you are in an airplane and your flight has reached cruising altitude, you can sometimes see frost forming in your cabin window. Jets cruise the lower levels of the stratosphere, which goes up quite a way. In fact, the stratosphere includes the 18-mile "it's beginning to look like space to me" limit, and then also extends another 13 miles to the edge of the mesosphere. When you see a bright falling star, it's burning up in the upper stratosphere.

THE OZONE LAYER AND BEYOND

The ozone layer occupies the lower portion of the stratosphere, generally between 12 and 19 miles above Earth, depending on the time of year. Not only does it provide protection from ultraviolet light, but it also provides a thermal lid to the troposphere.

So what happens in the ozone layer? When UV light strikes oxygen molecules in the lower stratosphere, it converts some of them into ozone molecules — a kind of hybrid oxygen, with three molecules instead of two. (Lightning strikes also produce ozone, which is why you can smell it in the air after storms.) These ozone molecules then absorb even more UV radiation from the sunlight, which splits them back to plain oxygen and releases a bit of heat at the same time. This cycle, called the Chapman cycle, is continuous, with oxygen molecules reacting with UV to create ozone molecules that split into oxygen molecules. The ozone layer is therefore considerably warmer than the air below and above it, sitting at about 0°C.

It acts as a thermal barrier, an inversion layer if you like, separating the frigid stratosphere from the equally frigid troposphere. That means that the ozone layer is the highest point that atmospheric convection currents can reach, and because convection is the engine of weather, driving everything from evening zephyrs to hurricanes, there is no weather in the stratosphere. The ozone layer's double duty is to protect life from unmitigated UV radiation and to put a vertical limit on weather.

The degradation of the ozone layer, which began in the late twentieth century, had such dire consequences for life on the planet that, in an unprecedented act of global cooperation, the world immediately banned the fluorocarbons that were destroying the ozone layer. But dangerously high levels of UV were not all that meteorologists were worried about. There was also the unknown consequences of losing the lid of the troposphere. What would have happened to climate worldwide? Hopefully we'll never know.

From the ozone layer to the top of the stratosphere is another 14 miles. (The stratosphere in total is about 20 miles thick.) The temperature gradually increases with height, so that by the time you get to the top of the stratosphere and the beginning of the next layer, the mesosphere, the temperature has reached a relatively warm -3°C. The mesosphere is a remnant atmosphere, more like a sprinkling of atoms, that extends from the top of the stratosphere, 31 miles above the Earth's surface, to 50 miles in altitude. All in all, it is about 19 miles thick, and it had to wait until the early 1960s before explorers reached it.

On July 17, 1962, almost exactly a century after Glaisher and Coxwell's precarious excursion to the beginning of the stratosphere, test pilot Robert M. White made a flight that put him on the cover of *Life* magazine and into the record books. White was a top gun. At age 20, he flew P-51 Mustangs over Germany in the closing stages of the Second World War. He was shot down, spent a year in a German prisoner-of-war camp and then, after being liberated, went right back to flying the fastest, hottest fighters the airforce could throw at him.

On that July morning, he was the pilot chosen to climb into the cockpit of an X-15 rocket slung under the wing of a flying fortress B-52. The B-52 took White to an elevation of 45,000 feet and released him. After a moment of freefall, he lit the rocket engine and tipped the nose of the X-15 almost straight up. "The horizon disappeared and I could see nothing ahead of me but sky," he later recounted. The G-force slammed him back in his seat, and, with the rocket engines on full, he watched the altimeter climb until he was at 120,000 feet. Then the engines cut off, their fuel exhausted. His X-15 was now ascending under pure momentum. In seconds, he was at 217,000 feet, his old record, but the X-15 was still climbing. "There was very little pressure on the control surfaces. The aircraft responded lazily," he reported. At that altitude, there is definitely not enough atmosphere to steer an aircraft, so his X-15 had been fitted with thrusters to adjust pitch and yaw. He was still climbing. He couldn't believe the altimeter — 300,000 feet, 310,000 feet, then finally 314,750 feet, a full 59 miles above the Earth's surface. White had flown through the entire mesosphere and on into the thermosphere above it.

In an interview after the flight, he recalled the experience: "But most impressive was the color of the sky . . . it's a very deep blue — not a night blue, just a deep, deep blue that's hard to describe but marvellous to see." And then the view beneath: "Wow! The Earth is really round . . . Looking to my left I felt I could spit into the Gulf of California; looking to my right I felt I could toss a dime into San Francisco Bay."

White was the first person to fly a fixed-wing aircraft into outer space. Yes, Yuri Gagarin and Alan Shepard had rocketed into space the year before, but they were passengers, not pilots. Everyone called them astronauts. The U.S. Air Force decrees that 50 miles high is the official, aeronautical edge of space, and any pilot who has flown above this limit is given official astronaut status by the U.S. Air Force and a medal to prove it. So White got his wings and the bragging rights to call himself an astronaut. (The Europeans set the bar higher. According to the Fédération Aéronautique Internationale, outer space lies on the other side of the Kármán line, 100 kilometers, or 62.1 miles, above our heads. White almost qualified for that designation too, though he wouldn't have received any hardware; the Fédération Aéronautique Internationale doesn't issue medals.)

The mesosphere is the realm, along with high-flying X-15s, of the ethereal, rare noctilucent clouds, made of meteor dust that can be seen in the summer months at higher latitudes. Beautiful but not really clouds at all. And if there is no weather in the stratosphere, there is certainly none in the mesosphere. At the mesosphere, meteorologists bail out. What goes on there doesn't really affect them. This is where the next tier of scientists kicks in — the exo-physicists and high-altitude researchers.

Scientists love dividing ambiguous phenomena into discrete slices, and atmospheric researchers are no exceptions. Many have made their careers by defining and extending ever more tenuous layers of atmosphere. At about 50 miles above the Earth's surface, the mesosphere gives way to the thermosphere, which stretches another 160 miles above the Earth's surface. The thermosphere is very hot — 500 to 2,000°C, though that is almost an abstract reading. There are too few particles per square yard to transfer any of that heat to objects moving through it. It couldn't melt a snowflake. Otherwise the space station, which circles our planet well inside the thermosphere on an orbit between 200 and 240 miles above the Earth, would burn up in a few seconds.

On the edge of space, the base of the thermosphere is the stage for Earth's greatest light shows: the aurora borealis (northern lights) and its southern cousin, the aurora australis. The northern lights are the most famous of the two because Canada, the Scandinavian countries and Russia are much closer to the North Pole than Australia, Chile, Argentina and South Africa are to the South Pole and so are seen by more people. From the vantage of space, the auroras crown each pole with electric iridescence, their rays stretching up from the mesosphere and through the thermosphere. They are a multihued electric particle show fueled by high-energy electrons from solar flares that have been caught by the Earth's magnetic field and corralled toward the poles. The eerie colors happen when these electrons collide with the sparse atmospheric atoms found at such high altitudes. Red and green hues are created by collisions with oxygen atoms, and the rarer violet hues are the result of solar electrons hitting nitrogen atoms. The shifting undulations in the "curtains" are caused by wavefronts of electrons moving at near light speed.

My father once told me that when he was a boy in Northern Ontario, he could hear, on very cold, still winter nights, the faint sound of the northern lights, a sort of dry, swishing sound. Since then I've talked to many a northerner who've said likewise, that the auroras sometimes whisper. I've also heard an old trapper say that if you whistle, the lights will respond by changing shapes. He said that if you were lucky you could even "call them down" and they will form above you. I'm not sure about that. A human whistle, even a loud one, can barely be heard more than a mile away, and auroras are more than 40 miles above the Earth's surface. Besides, there's not enough atmosphere up there to carry sound waves. Nevertheless, I do like the magical, intimate notion of communicating with these great, cosmic light shows as if they were some sort of entity.

A recent discovery has determined that the auroras have a younger brother called Steve (for strong thermal emission velocity enhancement). Discovered by aurora chasers in Alberta in 2017, Steve appears as a slightly curved vertical ribbon of white light in the sky that sometimes accompanies the northern lights. The ribbon is actually a stream of wildly hot gases (3,000°C), flowing at a speed of 770 miles per hour. Of course, Steve has been there all along; it's just that new high-resolution night photography has teased him out of the background.

These extravaganzas take place in a surprisingly rarefied atmosphere, like the vacuum inside a fluorescent light. Hundreds of miles above the Earth's surface, the few remaining atoms make up a gas so fleetingly insubstantial and diaphanous that it evaporates into outer space. This phenomenon is called Jeans escape, after the English astronomer James Jeans who first predicted this process. (The escape of atmosphere, mostly in the form of hydrogen atoms, is a negligible trickle now, but as the sun gets brighter — the sun brightens by 10 percent every billion years — the process will accelerate until the atmosphere disappears.) Once the atoms are in space, they are subject to solar wind, a force that has no effect whatsoever at sea level due to the protection of our atmosphere. Indeed, at such heights and with sparse populations, individual atoms of air get blown into space beyond the reach of Earth's gravity. They become interplanetary travelers bound for the farther reaches of the solar system.

You'd think the thermosphere would be the final edge of the atmosphere, but you'd be wrong. It seems scientists just don't know when to stop. They've added a final layer, the exosphere, which goes from the top of the thermosphere, 430 miles above Earth, to 6,200 miles into space. There are so few atmospheric particles in this region that they rarely collide. We wouldn't even know about the exosphere, but some overzealous engineer designed a particle detector so sensitive it could detect gas particles in parts per billion. Scientists found room for his device in a satellite bound for another planet. It detected occasional atmospheric particles up to 6,200 miles away from Earth. So the exosphere came to be. It's a little like calling the parking lot beside a baseball stadium the outer outfield, but no one's ever going to hit a ball that far.

3
CLOUD NINE
INSIDE THE MISTY GIANTS ABOVE OUR HEADS

"Cloud nine gets all the publicity, but cloud eight is actually cheaper, less crowded and has a better view."
GEORGE CARLIN

Clouds *are* weather. Without clouds, there would be no rain, snow, hail, sleet, tornadoes, hurricanes, typhoons, monsoons, lightning, floods, fog or rainbows. What would be left? Just wind and temperature. In a cloudless world, the evening news forecaster would be reduced to sunrise and sunset times, daily highs and nightly lows capped off with wind speed and direction. If, in a world without clouds, there was any moisture in the air at all, dew would be the sole form of precipitation. The only silver lining for news forecasters with a taste for theater would be dust storms, footage of oversize dust devils and extreme daily temperature variations. A cloudless world would be a dry world. Think the Sahara, where daytime highs of 38°C can be followed by nighttime lows of 0°C.

Insubstantial factories of infinite forms, clouds are both ephemeral and powerful, and though conjured out of practically nothing, out of ungraspable mist, they can shake the earth with thunder if they have a mind to. Like Joni Mitchell, those of us in the developed world who have flown have seen clouds from both sides, but unlike Joni claims in her song we *do* know what they are. For one thing, we know they're made of water vapor. But not many people are aware that the water vapor that makes up clouds is not like the mist from a sprayer nozzle or steam from a teakettle. Each droplet of water in a cloud is much, much smaller. It is only a millionth of a millimeter in diameter. Millions would fit into the period at the end of this sentence, and when billions upon billions of them congregate, they build the huge, fantastic shapes we call clouds. That's also why a typical cloud — a puffy, small fair-weather cumulus like the one William Wordsworth wrote about in his poem "I Wandered Lonely as a Cloud" — measuring a few hundred yards cubed, contains only a bathtub's worth of water.

Yet cloud droplets aren't the smallest form of aerosolized water. They coalesce from even smaller, airborne molecules of evaporated water much less than a millionth of a millimeter in diameter. The warmer the air, the more evaporated water it can hold; the cooler the air, the less evaporated water it can hold. As a matter of fact, there comes a limit, the dew point, when relative humidity levels reach 100 percent, or complete saturation. Only then can cloud droplets form. (Well, actually, there's one more necessity: cloud droplets need microscopic particles of dust to seed them — without dust, there would be no clouds.) The dew point can be hundreds to thousands of feet above the ground. Or it can be just a few inches. Fog is a cloud creeping on its belly.

You can see the dew point and relative humidity in action when you watch the contrail of a high-flying jet. They are literally seeding clouds, providing the microscopic particles necessary for water vapor to cling to. Some days, if the relative humidity is low, the contrails evaporate instantly. On other days, when the relative humidity is higher, they linger in long, tubular clouds. They form at that magic juncture of water vapor saturation and temperature, and in the constantly changing atmosphere every layer, every region, has a distinctive level of humidity. Cumulus clouds illustrate all of this.

ANATOMY OF A CLOUD

Here at the surface of the Earth, we live in a narrow layer of warmth, like minnows in the shallows of a summer lake. Passengers in hot air balloons know this well — the higher they rise, the colder it is. Even on a hot summer day, balloonists with just a few thousand feet under their belts begin to feel the distinct chill of altitude. Ten miles up, the temperature never rises above -40°C; that's why blankets are stowed aboard ballons and airliners..

This vertical decrease of temperature is called the lapse rate, and it's caused by two factors: thinner air holds less heat, and the higher you rise, the farther you are from the radiant heat of the ground, which is warmed by sunlight. On average, the temperature drops 3°C for every 1,000 feet of altitude. There's a difference of approximately 18°C between the base and peak of a 9,000-foot mountain. A warm spring afternoon in the valley translates into a freezing afternoon at the summit, which is why high mountains have snow on them all year round. It follows that most cloud bases, which start at 3,280 feet, must already be fairly cool.

Yet there's a sleight of hand, a sort of cover-up going on with the clouds. A cumulonimbus cloud, like the ideal one in the animated logo at the beginning of United Artists films, is an excellent example. Stretching from just above the Earth all the way to the stratosphere, a single cumulonimbus spans three critical thresholds. The dew point is the flat base of the cloud. Below that line, the air is saturated but not yet to 100 percent. Above that line, called the lifting condensation level, it is. You could take a seven-year-old by the hand and point and say, "See where the bottom of the cloud forms? That's the dew point." It's so clear.

What's not obvious is the freezing line. Let's imagine that our cumulonimbus cloud has formed on a hot summer day with a ground temperature of 30°C. If we apply the lapse-rate formula, then the elevation at which the temperature drops below freezing is approximately 16,400 feet above the ground. And here's the riddle: why isn't there any demarcation that indicates where a cloud goes from a liquid to a solid? Seeing a whole eight-mile-high cumulonimbus from several miles away, you'd think you would see a line, like the dew point, or perhaps a color difference, but there's nothing. And that's because there *is* none. There's certainly a threshold above and below which precipitation — rain or snow or hail — is either frozen or liquid, but not the fabric of the cloud itself. Why? It all boils down to surface tension, the "skin" that forms between water and air. The smaller the body of water, the greater the surface tension. A dime-sized puddle of water on the floor has rounded edges, while a bead of dew on a leaf is almost spherical. The tiny water droplets that form clouds are so small that their high surface tension prevents them from freezing. Cloud droplets remain liquid right down to -40°C. Only then do they freeze. (Unless they're sprinkled with a little silver iodide . . . but more on that later.)

This subthermal property of cloud droplets explains why the dew point has the same effect under subzero conditions. Cirrus clouds, at 18,000 feet and above, undergo their whole life cycle, from formation to evaporation, in subzero temperatures. In fact, when a grand cumulonimbus turns into a storm cloud and builds up until it reaches the bottom edge of the stratosphere, some 42,240 to 52,800 feet up, it smears out sideways, like smoke trapped under a glass ceiling. This is where the top of the troposphere meets the lower stratosphere, neatly delineated for us. Here the shearing winds of the lower stratosphere flatten and stretch the top of the cumulonimbus into a long cirrus cloud of frozen vapor. From a distance, the profile of the entire cloud resembles an anvil, hence the name cumulonimbus incus (*incus* is Latin for anvil). It's the ultimate storm cloud. Inside, it's pure bedlam.

GETTING A HANDLE ON CLOUDS

The ancient Greeks always seem to be the first to propose anything like a rational explanation for natural phenomena, and, as always, without scientific instruments, their speculations were guided by logic and observation. One of the earliest Greeks to wrestle with the enigma of clouds was Anaximander (610–546 BCE). He proposed that lightning was a product of friction within clouds (which, in a way, it is, if you think of it as analogous to the accumulation of static electricity on the surface of a balloon rubbed over someone's hair) and that wind was "a flowing of air," being moved by unknown forces. Two for two.

Later, early in the fourth century BCE, the philosopher Theophrastus from the island of Lesbos thought that the shape of clouds could be used to forecast weather. He observed, "If in fair weather a thin cloud appears stretched in length and feathery, the winter will not end yet," and, "Fear not as much a cloud from the land as from ocean in winter; but in summer a cloud from a darkling coast is a warning."

It was Democritus (460–370 BCE) who, along with his prescient theory of atomic particles, first speculated on how clouds were formed. He wrote that the vapors from melting snow and ice (he had traveled to northern Europe) were carried aloft by the wind southward until they hovered over the lakes that fed the Nile, then they fell as rain. This extraordinary speculation not only correctly described the water cycle of evaporation, recondensation and rainfall but also hinted at the vaster systems of prevailing winds and seasonal weather patterns.

Aristotle (384–322 BCE) was the first philosopher-scientist to write a meteorological treatise. It was based on his broader theories that the world was in constant flux and everything we see, everything around us, boiled down to the interplay between earth, air, fire and water — with earth at the center surrounded by water, then air, and then the outer sphere of fire. This is why, at the beginning of his meteorological musings, he wrote that the key question about clouds is why do they "not form in the upper air as one might on the face of it expect?" He answered the question by pointing out that the sun's heat reflecting off the ground prevents clouds from forming there, and that they do not form too high in the atmosphere because there is too much fire there. Thus, clouds form in the happy medium zone — the middle atmosphere — between two hot spots. It's a logical speculation, though if he'd known about the lapse rate he'd have had to change his theory.

The Romans tackled clouds as well. One of the earliest was Lucretius (99–55 BCE) who, with startling insight, suggested that clouds began as a "sudden coalescence, in the upper reaches of the sky, of many flying atoms of relatively rough material, such that even a slight entanglement clasps them firmly together." It was as close to the concept of the dew point, where water vapor in the form of humidity condenses into water droplets around particles of dust, as any thinker in antiquity got.

Seneca (4 BCE–65 CE), a great observer of Roman society, also wrote about weather and clouds. In his *Natural Questions*, a 10-volume treatise on natural history, Seneca focused on the changeability of the atmosphere and weather: "now rainy, now clear, now a varied mixture between the two. Clouds — which are closely associated with the atmosphere, into which the atmosphere congeals and from which it is dissolved — sometimes gather, sometimes disperse, and sometimes remain motionless." He correctly viewed clouds as indivisible from the weather.

The famous Roman natural historian Pliny the Elder (23–79 CE) wrote about weather around the same time, iterating Democritus's view of the atmosphere playing host to a great cycle of water vapor rising and falling. "Steam falls from on high and again returns on high," which, in essence, is another very concise description of the water cycle. He was on the right track, but it took almost two millennia before a more scientific view of clouds, and cloud formation, was put forward during the Renaissance by the French philosopher René Descartes. In his *Discourse on Method* (1637), Descartes proposed a methodology that, if followed correctly, would lead to the complete understanding of any natural phenomenon. Clouds became an example for him.

He wrote that humankind had always mythologized clouds; that because of their lofty abode, they had come to be associated with deities, and Descartes decided it was time to bring them down to earth. The elusive clouds were a natural target for his rationalism; they were so changeable, so ephemeral, so seemingly unclassifiable.

He got a lot of the science right. He speculated that clouds were composed of water droplets or small particles of ice that coalesced into "little heaps, and these gather'd together compose vast Bulks." They stay aloft because they are "so loose and spungy, that they cannot by their weight overcome the Resistance of the Air." But if they keep coalescing, he went on to say, they wouldn't be able to resist the pull of gravity and would then fall as rain or snow. Except for a few particulars, he was on the money.

Almost 200 years later, Luke Howard (1772–1864) turned his attention to meteorology. As a 16-year-old living with his Quaker family in Stamford Hill (now a part of North London), he built in his back garden a weather observation station consisting of a thermometer, a rain gauge and a recording barometer. He took readings twice a day and kept records in a journal, listing wind direction, air pressure, rainfall and maximum and minimum temperatures. Four years later, he was working as a chemist and apothecary in partnership with William Allen, the founder of a scientific club called the Askesian Society.

Public interest in science was burgeoning, and England was chockablock with gentlemen's clubs and learned societies, mostly populated by well-heeled amateurs, many of them natural historians. The Askesian Society was perhaps one of the most eccentric of these societies. Like other clubs of the period, its members attended lectures delivered by fellow members and guests. These usually consisted of the reading of learned papers, as well as popular experiments, often a showy demonstration of a chemical reaction.

Of all the phenomena the scientific world presented to the Askesian Society, the members were most curious about psychoactive substances, particularly nitrous oxide. William Allen claimed it had "a remarkably inebriating effect." One of the popularizers of nitrous oxide (now known to us as laughing gas), Humphry Davy, became so addicted to "the gaseous oxide of azote," as it was then exotically referred to, he confessed to taking it "three or four times a day." So it was a jolly group of neophyte scientists. Very unlike a Quaker congregation.

Luke Howard had tried nitrous oxide along with the other members of the Askesian Society, but his central fascination was meteorology. Howard wished to do for clouds what Carl Linnaeus (1707–1778), the Swedish wunderkind naturalist, had done for zoology: namely to develop an overarching classification system for all cloud types.

It was only natural for Howard to emulate Linnaeus. He was universally regarded as one of the greatest scientists of the eighteenth century. Linnaeus's *Systema Naturae*, published in 1735, was an astounding advancement in zoological nomenclature. In it, he introduced a binomial, Latinate system of describing species that we still use today. We humans, for example, are famously *Homo sapiens*, where *homo* is the genus and *sapiens* is the subspecies. The Linnaean classification system is so comprehensive that every living thing, and everything that has ever lived, including all fossils, has a place within in it. Even undiscovered species will have a place in the system. Without Linnaeus, Darwin couldn't even have begun his theory of evolution. Johann Wolfgang von Goethe once said of Linnaeus, "With the exception of Shakespeare and Spinoza, I know no one among the no longer living who has influenced me more strongly."

Luke Howard's personal life, a marriage and various apprenticeships as a chemist, had slowed his meteorological investigations, and it wasn't until fate relocated him to Plaistow, a suburb on the outskirts of London, that he was again able to resume, as he later wrote, "the observations I had long been making on the face of the sky." With no more equipment than a good eye and a crick in his neck, Howard set out to classify the unclassifiable: the ineffable, vaporous, ever-changing clouds. He could see an order to them, one that after a few more years of keen observation and note-taking, was robust enough to publish as a scientific paper. So in December 1802, Luke Howard, a member in good standing, read his paper entitled *On the Modifications of Clouds* to the Askesian Society. It changed meteorology forever.

Howard set out three basic cloud types: cumulus, stratus and cirrus. But there were combinations of these, which added up to 10 types: cirrus, cirro-cumulus, cirro-stratus, alto-cumulus, alto-stratus, cumulus, strato-cumulus, nimbo-stratus, stratus and cumulo-nimbus. It was breathtakingly simple yet completely inclusive. He had pulled off a Linnaean coup for clouds. Shortly after his paper was delivered, he became a scientific celebrity across Europe. Even Goethe was impressed, writing, years later in 1817, this short appreciative doggerel:

> *But Howard gives us with his clearer mind*
> *The gain of lessons new to all mankind;*
> *That which no hand can reach, no hand can clasp*
> *He first has gained, first held with mental grasp.*
> *Defin'd the doubtful, fix'd its limit-line,*
> *And named it fitly. — Be the honour thine!*
> *As clouds ascend, are folded, scatter, fall,*
> *Let the world think of thee who taught it all.*

CLOUD WATCHING

In 1887, the first *International Cloud Atlas* was published. It was replete with illustrations of Howard's cloud types and some of the first published color photographs. We know today that these various types of clouds, almost like separate species, inhabit different altitudes, and any of us who have flown through and above them are cloud explorers. We have felt the turbulence of updrafts within cumulus clouds and seen panoramas no earthbound observer has ever witnessed. For a cloud-gazer like me, a nephological aficionado, the wonder of clouds outside an airplane window never fails to astonish and inspire.

A descending airliner samples Howard's cloud almost in the same order as the *International Cloud Atlas*, starting with cirrus clouds (Latin for "lock of hair"), which hover at altitudes of 16,500 to 50,000 feet (3 to 9 miles) above the ground. These lofty clouds have the most subspecies, 13 in all, including cirrus incinus ("mares' tails") and cirrocumulus stratiformes (the "mackerel sky" that often spreads from horizon to horizon). Every one of the 13 are entirely composed of ice crystals. Often cirrus clouds have the look of brushstrokes or tufts of hair. I have raced by cirrus incinus on airplane flights near cruising altitude, their wispy tails flickering so close I could have reached out to touch them.

Stratus clouds (Latin for "layer") can be as high as cirrus clouds or as low as cumulus clouds. They occur at the boundaries of different air temperatures and are flat and layered, because no vertical convection columns puff them up. They have seven subspecies among which is altostratus, forming just below cirrus altitude. Altostratus are often semitransparent, so you can see the disc of the sun through them. I remember one flight where, as we began our descent, we appeared to be landing on a vast field of stratiform clouds that rose up over the sides of the airplane like mist over a lake, a cotton fog we sank into. The blue daylight dimmed until, deep inside the cloud, the windows were a dark pearl gray, as if they'd all been painted. Then suddenly, the windows brightened, and we came out of the bottom of the stratus layer into a gray, twilight world sandwiched between two layers of clouds — altostratus above and nimbostratus below. (*Nimbus* is the Latin word for rain.) It was a parallel universe in which somber cloud vistas devised strange landforms to the horizon — low gray mountains and great plains. Then our flight sank into this nacreous landscape without a tremor, submerging again into the pearly darkness.

Classic cumulus clouds (Latin for "heaped") are generally low clouds, and their bases often hover less than 6,500 feet above the ground. From above, cumulous clouds seem like puffy mountains with no features, no rivers or forests, only snow-white landscapes. The unbroken whiteness makes them seem heavenly, like celestial terrain transformed, purified.

Cumuli are the mascots of the empyrean vault of heaven, endless form seemingly without purpose or function, sheer natural invention. Stratus clouds enclose relatively little turbulence, but cumulous clouds always contain updrafts and so the ride through is usually a little bumpy. Being a bit of a backseat pilot, I've always felt a sense of relief as my flight passes out of a cumulus cloud.

The 1896 edition of the *International Cloud Atlas* stood for more than 50 years; the edition that meteorologists use today was published in 1951. This atlas used what was referred to as the C code and started at 0 instead of 1. It also did away with the cumbersome hyphens. The order of the cloud types is now as follows: 0. cirrus, 1. cirrostratus, 2. cirrocumulus, 3. altocumulus, 4. altostratus, 5. nimbostratus, 6. stratocumulus, 7. stratus, 8. cumulus, 9. cumulonimbus. Here, finally, is the origin of the term "cloud nine."

The tallest, grandest cloud of all — in extreme cases reaching heights of 12 to 15 miles, more than twice the height of Mount Everest — closes out the list of cloud types, and of all the clouds it has the most complex and dynamic inner life. When one of these monster cumulonimbus clouds rises until it hits the top of the troposphere, it acquires the classic anvil shape and now takes on one final Latin moniker: cumulonimbus incus. It is a weather system unto itself. No one has explored the interior of one of these clouds by balloon or purposely by airplane; they are just too dangerous. But one American became a pioneer through circumstances beyond his control.

During the late afternoon of July 26, 1959, U.S. Marine Corps Lieutenant Colonel William Rankin and his wingman, Lieutenant Herbert Nolan, were piloting a pair of F-8 Crusader fighter jets to a base in Beaufort, North Carolina. They were flying at an altitude of 47,000 feet (9 miles) to avoid turbulent weather beneath them. Just before they began their descent, Rankin heard grinding noises coming from his engine. Then his instrument lights went out, and when he pulled the emergency auxiliary power lever, it broke off in his hand. He radioed Nolan and told him he was ejecting.

At this altitude, the ambient temperature is -50°C. There is next to no oxygen and the air pressure is less than a third of that at the surface. Rankin wasn't wearing a pressure suit. As soon as his canopy blew off and his chair rocketed out of the aircraft, his body began to decompress. His ears, nose and mouth started bleeding and his stomach became dangerously distended. His glove had been ripped off and he could feel frostbite instantly numbing his hand. Fortunately, he had an oxygen canister attached to his helmet, so he remained conscious.

His descent — from ejection to touchdown — should have taken about eight minutes, a 3.5-minute freefall to 10,000 feet where his parachute would automatically deploy. But he had bailed out over a thunderstorm. Violent updrafts within the cumulonimbus cloud held him in their grip for almost 40 minutes.

> *I was blown up and down as much as 6,000 feet at a time. It went on for a long time, like being on a very fast elevator, with strong blasts of compressed air hitting you. Once when a violent blast of air sent me careening up into the chute I could feel the cold, wet nylon collapsing about me. I was sure the chute would never blossom again. But, by some miracle I fell back and the chute did recover its billow.*
>
> *The first clap of thunder came as a deafening explosion that literally shook my teeth. I didn't hear the thunder, I actually felt it — an almost unbearable physical experience. If it had not been for my helmet, the explosions might have shattered my eardrums.*
>
> *I saw lightning all around me in every shape imaginable. When very close, it appeared mainly as a huge, bluish sheet several feet thick. It was raining so torrentially that I thought I would drown in midair.*

Several times I held my breath, fearing that otherwise I might inhale quarts of water.

Rankin eventually landed in a forest and made his way to a road where he flagged down a passing motorist. Later, in hospital, he was treated for bruises, frostbite and severe decompression.

This is why airliners avoid flying through active cumulonimbus clouds. Lightning is scary up close but it seldom strikes an airplane, and heavy rain just blows through the fans of jet engines. It's those updrafts and downdrafts — particularly if they're side by side. They can, and have, torn airliners apart. So cloud nine is hardly the bucolic, tranquil paradise it has been made out to be. Just ask William Rankin.

4
THE POEM OF EARTH
RAIN

"I feel sorry for short people, you know.
When it rains, they're the last to know."
RODNEY DANGERFIELD

A cloud is a galaxy of microscopic water droplets. Immune to gravity, the droplets float suspended in the air like bubbles in syrup. In most clouds, this weightless multitude exists in a steady state, but if the cloud is large and there are other forces acting on the droplets — thermal convection updrafts and downdrafts for one, or a change in altitude as a cloud is pushed up and over a mountain — then some of the droplets collide and combine into larger droplets. If they keep coalescing, they eventually get so large that gravity begins to pull them downward, collecting even more droplets until they become raindrops. When vapor coalesces higher in a cloud, as growing crystals of snow, these too will begin to fall, but they'll melt into rain if the temperature at the base of the cloud is above freezing.

If we follow one raindrop as it falls from the cloudy heights, we'd have to keep quickening our pace because as it falls it picks up speed. To a point. In a vacuum, the raindrop's acceleration would continue indefinitely, but air resistance slows its fall. When the forces of gravity and air resistance balance each other, the drop ceases to accelerate and reaches its terminal velocity. Of course, the size and weight of the raindrop decides what the velocity will be: the heavier, the faster. Not that rain contradicts Galileo's rule — large objects accelerate at exactly the same rate as small objects — it's just that smaller drops of rain encounter more air resistance as a ratio of their size to mass. It's the same reason a feather falls more slowly than a pebble. Drops of drizzle, which are less than 0.5 millimeters across (salt-grain size), have a terminal velocity of 4.5 miles per hour, while a large raindrop about five millimeters across (house-fly size) falls at the rate of 20 miles per hour. By comparison, a falling human being hurtles to the ground at a terminal velocity of about 125 miles per hour.

Raindrops are not teardrop shaped. The smallest, like those that make up drizzle or Scotch mist, are almost perfectly spherical. As they get larger, into the five millimeter range, their bottoms flatten out with air resistance and they assume a sort of bun-like profile. Raindrops larger than five millimeters get a dimple indent in the bottom of their buns and begin to look more and more like mushroom caps or fat parachutes. Nine millimeters is the upper limit for raindrop size. Any larger than that and they break up into smaller drops because, at higher terminal speeds, air resistance increases by the square of the velocity.

Another factor affecting the speed of the fall is the density of the atmosphere. A raindrop on Mars would fall faster than one on Earth, even with lower gravity, because there is very little air resistance. Our own planet has probably had variations in the density of its atmosphere over the millennia, but we have had no way of measuring those fluctuations until recently, when fossils of raindrop impressions were discovered on a farm near Prieska, South Africa. A layer of fresh ash from a volcano preserved these traces as the ash transformed into rock. When the rock was dated, it turned out that this passing shower occurred 2.7 billion years ago. Scientists analyzing the tiny impact craters realized that they were more than a time capsule, they were a snapshot of the thickness of atmosphere during the Great Oxygenation Event. They estimated that the ancient raindrops measured 3.8 to 5.3 millimeters across and that, given the radius of the splashes, the atmospheric density was not that much different from today. So the wind on our time traveler's face would feel the same as any wind on Earth today.

APHORISTIC SHOWERS

"Moods light up the world."
MARTIN HEIDEGGER

"Lovely day, isn't it?" my neighbor said this morning. I had to agree — sunny, warm, with just the hint of a breeze. She gestured at the sky. I looked up and noticed that, except for a few irregular shreds of cumulus fractus clouds, which are barely clouds at all, the autumn sky was unbroken blue. She was smiling, and I have to admit my own mood was so buoyant that finding a utility bill in my mailbox didn't phase me at all. A sunny day lifts everyone's spirits. As Marcel Proust wrote in *Remembrance of Things Past*, "A change in the weather is sufficient to recreate the world and ourselves."

The intimate relationship between weather and mood plays out in common sayings. To comfort someone who is *under the weather*, we offer them *a ray of hope* and reassure them that *every cloud has a silver lining*. An easy task is *a breeze*, but you can be *snowed under* by a difficult one. Furthermore, it seems we all contain our own internal weather. An optimist has a *sunny disposition*; a dreamer walks with her *head in the clouds*. Someone with anger management problems has a *stormy* temperament, and a *fair-weather friend* is fickle, definitely not *right as rain*. A lull before a conflict is *the calm before the storm*. Rainy days can bring on the blues: "Rainy day go away, come again some other day."

It's hard to tell sometimes whether the weather is getting you down or if you're just down. Seasonal affective disorder is definitely a case of the former, where the sufferer's mood declines with the waning daylight hours of northern latitudes. People afflicted with migraines report that rainy, low-pressure systems bring on their worst headaches. But can the reverse be true also? Can our inner moods, by some subtle telekinesis, affect the weather? Being vainglorious creatures we sometimes think the weather is mirroring our emotions. In literature, this is called pathetic fallacy — the wind is angry, those gray clouds are sullen.

LITERARY RAIN

Writers tease out the intimate implications of weather as experienced by emotional, sentient beings. In the sensorium of experience, the greatest writers have always been masters of inflection and analogy. Their summaries of phenomena are as precise, in their way, as the most detailed observations of science. And literature is replete with rainy days.

It's safe to say that gray, rainy days are moody or somber, but rainy nights can be gloomier still. Regrets and lost loves can haunt a poet like Edna St. Vincent Millay on a rainy night. Consider the beginning of her "Sonnet XLIII."

What lips my lips have kissed, and where, and why,
I have forgotten, and what arms have lain
Under my head till morning; but the rain
Is full of ghosts tonight, that tap and sigh
Upon the glass and listen for reply,
And in my heart there stirs a quiet pain
For unremembered lads that not again
Will turn to me at midnight with a cry.

Ray Bradbury wrote perhaps the bleakest passage about nocturnal rain in his novel *Green Shadows, White Whale*: "I went to bed and woke up in the middle of the night thinking I heard someone cry, thinking I myself was weeping, and I felt my face and it was dry. Then I looked at the window and thought: Why, yes, it's just the rain, the rain, always the rain, and turned over, sadder still, and fumbled about for my dripping sleep and tried to slip it back on." Bradbury was writing about a rainy night in Ireland, which has almost as much rain as England. Talk about dismal. If his character had lived in London, England, during the late Victorian era, his mood might have been even worse. For the entire month of December 1890, the London weather bureau didn't register a single minute of sunlight.

Thomas Merton wrote about a mystical experience he had listening to the rain in darkness: "What a thing it is to sit absolutely alone, in the forest, at night, cherished by this wonderful, unintelligible, perfectly innocent speech, the most comforting speech in the world, the talk that rain makes by itself all over the ridges, and the talk of the watercourses everywhere in the hollows!"

That's more like it. Now we are in the poetic realm of the sublime and mysterious, and Merton is far from being the only writer to hear voices in the rain. In his novel *Still Life with Woodpecker*, Tom Robbins described the West Coast rain falling over an hallucinogenic landscape in Seattle: "The thin, gray rain that toadstools love. The persistent rain that knows every hidden entrance into collar and shopping bag. The quiet rain that can rust a tin roof without the tin roof making a sound in protest. The shamanic rain that feeds the imagination. The rain that seems actually a secret language, whispering, like the ecstasy of primitives, of the essence of things."

Pablo Neruda claimed that he absorbed the language of the rain when he was a boy: "my poetry was born between the hill and the river, / it took its voice from the rain." Walt Whitman not only listened to the rain speaking, he conversed with it. In his 1885 poem "The Voice of the Rain," Whitman asks "who art thou?" and the rain replies in a marvelous synthesis of lyrical and scientific veracity:

I am the Poem of Earth, said the voice of the rain,
Eternal I rise impalpable out of the land and the bottomless sea,
Upward to heaven, whence, vaguely form'd, altogether changed, and yet the same,
I descend to lave the drouths, atomies, dust-layers of the globe,
And all that in them without me were seeds only, latent, unborn:
And forever, by day and night, I give back life to my own origin, and make pure and beautify it.

Whitman picked up the theme of fertility in his poem, echoing the work of Aeschylus 2,000 years before him. In Aeschylus's *Danaids*, Aphrodite proclaims her dominion over the generative cycle:

The pure sky loves to violate the land,
and the land is seized by desire for this embrace;
the teeming rain from the sky

makes the earth fecund, so that for mortals it generates
the pastures for their flocks and the sap of Demeter
and the fruit on the trees. From these moist embraces
everything which is comes into being. And I am the
cause of this.

John Updike maintained the heavenly threnody in his novel *Rabbit Is Rich*: "Rain is grace; rain is the sky condescending to the earth: without rain, there would be no life."

Douglas Coupland grew up in rainy Vancouver, and I suspect that, unlike many others, he found gray rainy weather personally nourishing. A character in his collection of short stories, *Life After God*, confesses that "The richness of the rain made me feel safe and protected; I have always considered the rain to be healing — a blanket — the comfort of a friend. Without at least some rain in any given day, or at least a cloud or two on the horizon, I feel overwhelmed by the information of sunlight and yearn for the vital, muffling gift of falling water."

There is a special smell, a sort of intoxicating perfume that accompanies the rain, especially after a dry spell. Barbara Kingsolver described it in *The Bean Trees*: "That was when we smelled the rain. It was so strong it seemed like more than just a smell. When we stretched out our hands we could practically feel it rising up from the ground. I don't know how a person could ever describe that scent."

Decades before Kingsolver wrote that passage, two Australian mineralogists, Isabel Joy Bear and Richard Grenfell Thomas, had already not only set out to describe the scent of rain, but also to discover its earthen recipe. They ground up sunbaked rocks and clay and used steam distillation to extract the scent from the mixture. The terpenes that the rocks had absorbed from the atmosphere made up the bulk of this perfume. Bear and Thomas published their results in the scientific journal *Nature* in 1964, calling the scent "petrichor," from the Greek roots for "rock" and *ikhor*, the blood of the gods. Paradoxically, nothing was better at instilling the highest levels of this scent in Australian rock and clay than a protracted drought.

DROUGHT, RAINMAKERS AND PROVIDENCE

When the American explorers Lewis and Clark first encountered the Nebraska and Kansas territories in 1804, they noted in their journals that the region was too dry to support agriculture. Yet six decades later, a westward-flowing tide of settlers, encouraged by railroads and land-grab speculators, began to settle there. And once these pioneer farmers started to plow the fields, something extraordinary happened — a phenomenon that seemed to beg divine providence — it started to rain. Plentifully. The year 1870 heralded two decades of uncharacteristically high rainfall that oversaw the establishment and expansion of thousands of farms.

The hubris of manifest destiny was embodied in a new pioneer maxim: "rain follows the plow," and this saying was soon given quasi-scientific legitimacy. A professor at the newly established University of Nebraska, Samuel Aughey, explained the increased rainfall as a direct result of cultivation, which caused a "great increase in the absorptive power of the soil." This, he insisted, led to more moisture being captured by the soil and therefore more evaporation and more clouds and more rain, and so on. It was logical enough. For two decades, downpour after downpour, his theory held true and the land grew fertile. Then, in 1890, the rain stopped.

Within two years, there was a mass exodus. The drought-stricken region of western Kansas and Nebraska lost almost half its population. But the remaining homesteaders adapted. Well into the twentieth century, they eked out a hardscrabble existence, planting drought-tolerant crops like millet, sorghum and winter wheat. It wasn't bountiful, but if you were prepared to cut corners and scrimp and save, the land was sufficient.

The dearth of rain wasn't the only scourge the homesteaders faced. In the midst of their misery, a new kind of entrepreneur began to stalk the parched farming communities and small towns — the rainmaker. Frank Melbourne, popularly known as "the Rain Wizard" or "the Australian," was one of the most famous. He was tall and imposing, and he sported a dark beard that gave him an Old Testament authority. Born in Ireland, he had immigrated to Australia. This must have been where he learned to make rain, for he told his clients that he was forced to flee Australia to avoid arrest for causing floods. Hmm. Well, the farmers bought his story.

Melbourne's fee for a soaking rain, one that covered at least 50 square miles was $500, a tremendous sum in those days, especially for cash-strapped homesteaders. He worked farming communities from Canton, Ohio, all the way to Cheyenne, Wyoming. His technique was mysterious. It centered on his "Rain Mill," a machine that was said to utilize a crank and special gases, though no one had ever seen it. He carried it in a big black rucksack with his pistol at the ready to discourage overly curious onlookers. In Cheyenne, where he convinced 23 farmers to pool their savings, he locked himself in a stable and covered all the windows with blankets. There he worked his quiet magic.

It poured the next day, and the *Cheyenne Daily Sun* lauded his success in a glowing article. Later, in Goodland, Kansas, he operated the Rain Mill for several days without drawing a shower, though it did rain in other parts of Kansas. Melbourne claimed that the wind blew his influences off course. Even with that middling success, the *Chicago Tribune* wrote, "Melbourne Causes the Rain to Fall. Complete Success Attends his Latest Experiments at Goodland." Suddenly Melbourne was a hot ticket on the arid plains. He plied his trade with mixed success over the next few years and then, mysteriously in 1894, after selling his secret to three Goodland businessmen, he committed suicide in a Denver hotel room. Each of the businessmen went on to form a rainmaking company: the Goodland Artificial Rain Company, the Interstate Artificial Rain Company and the Swisher Rain Company.

Another Goodland citizen, a railway man by the name of Clayton B. Jewell, figured out Melbourne's method on his own. He successfully pitched his technique to the Rock Island Railway, and it underwrote and refurbished a special rainmaking rail car for his personal use. Though most of his apparatus was hidden inside, there were tubes projecting from the roof for releasing special gases. Inquisitive onlookers stared as vapors flooded out of the tubes and dissipated into the sky. He became even more popular than Melbourne. But soon, like a lot of other smart rainmakers of the time, he headed west to California. That was where the big money was.

Unfortunately, California was not an El Dorado for Jewell or for a dozen or so other rainmakers. Over a few dry years, he lost his reputation. But the rainmaking chapter had not yet closed. The most renowned rainmaker of all, Charles Hatfield, began his career just as the others were fading out.

Unlike his forebears, Hatfield was familiar with meteorology. As a boy in San Diego, he had read all the meteorology books in the municipal library and become seized with the idea that it might be possible to stimulate the formation of rain clouds with chemical vapors. In 1902, on a clear sunny day, he climbed the windmill on his father's ranch with a concoction he'd mixed in a shallow pan. Within a day, it rained.

Charles Hatfield had seen his destiny. With pale blue eyes, high cheekbones and his quasi-scientific vocabulary, he looked and sounded like a born rainmaker. In 1904, he made a bet with 30 Los Angeles businessmen that he could draw 18 inches of rain for Los Angeles over the winter and following spring of 1905. He referred to himself not as a rainmaker, but as a "moisture accelerator." "I cannot make it rain," he claimed. "I simply attract clouds and they do the rest." He built his rain derrick — a 20-foot wooden tower crowned with a metal tank — and began to draw the rain. Onlookers were always impressed watching him work. He would climb up the towers and stir the chemicals in the tanks sending great plumes of vapor billowing into the air.

The *Los Angeles Examiner* published an interview in which he explained his technique: "When it comes to my knowledge that there is a moisture-laden atmosphere hovering, say, over the Pacific, I immediately begin to attract the atmosphere with the assistance of my chemicals, basing my efforts on the scientific principle of cohesion. I do not fight Nature . . . I woo her by means of this subtle attraction."

Of course, it poured that winter, and Hatfield's fame grew. He began to get contracts up and down the West Coast, from British Columbia to Mexico. His only opposition came from the chief of the U.S. Weather Bureau, Willis Moore. Every time a newspaper published an article praising Hatfield's rainmaking abilities, Moore would make sure the weather bureau published a withering disclaimer. Even the San Diego weather bureau was threatening to charge Hatfield with fraud. But the public knew better: Hatfield could make it rain.

Perhaps it was hubris, or maybe it was the pride of a hometown boy trying to prove himself, but San Diego turned out to be his nemesis. A ruinous drought had parched the region for four long years, and the city reservoir was down to a third of its capacity. Hatfield, the homegrown hero, negotiated an all-or-nothing verbal contract with the San Diego city council — for $10,000, he would fill the reservoir by December 1916. They would owe him nothing if he failed.

Hatfield built an evaporating tower beside the reservoir and started producing his vapor on January 1. Five days later, it began to rain. Then the heavens opened. San Diego received 28 inches in one month. Not only did the reservoir fill and overflow, but on January 27, two nearby dams, the Sweetwater and the Lower Otay Lake, burst and flooded their valleys. Railways, bridges and roads were destroyed along with hundreds of homes. Twenty deaths were attributed to the floods, and rumors started to fly that angry citizens had formed an armed posse to lynch Hatfield. But he had already escaped the city on horseback.

Boldly, Hatfield returned in February to collect his fee, but the city council refused to pay him unless he accepted liability for the damages, a sum of $3.5 million. It wasn't his fault, he insisted; the city should have made adequate preparations for the deluge. It didn't help that there was no written contract. Hatfield persisted, first trying to settle for $4,000 and then, when that failed, he sued the council. He pressed his case unsuccessfully for 22 years. But in the two subsequent trials, the courts ruled the rain was an act of God.

DROUGHT

Hatfield continued to ply his trade until the stock market crash of 1929 brought an end to his career. In 1931, a record drought struck eastern North America; the following year the drought worsened, spreading westward to the Midwest and the Great Plains. Summer temperatures soared. During a heat wave in Illinois in 1934, 370 people died, and two years later, between July 5 and 17, 1936, an even deadlier heat wave struck Manitoba and Ontario, resulting in 1,180 deaths. The mercury peaked at 44°C, causing steel rail lines and girders to twist and buckle and the tar to melt on the roads. In the countryside, farmers watched helplessly as their fruit crops cooked on the trees.

The breadbasket of North America turned into the Dust Bowl. The earth dried and cracked open with great fissures deeper than the reach of a man's arm. One of the most dramatic consequences of the decade-long drought was a series of epic dust storms. In the Dirty Thirties, it was dust, not rain, that followed the plough. Dry, furrowed earth crumbled into sand and desiccated loam. Fields with failed crops had no roots to hold the soil, and so much land had been cleared of prairie grass that the dry winds could scoop up the topsoil and carry it high above the clouds. Texas, Oklahoma, Kansas, Colorado and New Mexico were particularly hard hit by big dust storms. "Dusters," as they came to be known.

Folk singer Woody Guthrie survived one of the worst of these in Texas when he was 23. The day the storm struck, April 14, 1935, is now remembered as Black Sunday. Guthrie and his neighbors watched a thousand-foot-high dust cloud roll in before they ran to their homes for cover as the howling blackness descended upon them. He wrote,

> *We sit there in a little old room, and it got so dark that you couldn't see your hand before your face, you couldn't see anybody in the room. You could turn on an electric light bulb, a good, strong electric light bulb in a little room and that electric light bulb hanging in the room looked just about like a cigarette burning. And that was all the light that you could get out of it.*
>
> *A lot of the people in the crowd that was religious-minded, and they was up pretty well on their scriptures, and they said, "Well boys, girls, friends, and relatives, this is the end. This is the end of the world." And everybody just said, "Well, so long, it's been good to know you."*

But this was not the worst of it. The biggest dust storm of all was a two-day marathon that had struck the American plains on May 9, 1934, a year earlier. It is estimated the storm picked up 350 million tons of topsoil, 12 million of which ended up being dumped on the Chicago region while the rest blew east, darkening the skies in New York, Washington and Boston before moving out to sea. Ships hundreds of miles from shore in the Atlantic Ocean encountered heavy dust falls. On some of their decks, the dust accumulation was a quarter-inch thick. By the end of the 1930s, 75 percent of the topsoil in the North American plains had blown away.

THE LAST RAINMAKER

The Second World War saw a return to normal rain levels and the gradual replenishment of the topsoil. Regional droughts have continued to plague California, but droughts on the scale of the 1930s have not returned. Rainmakers never returned either, except for one strange exception: Wilhelm Reich.

Reich was an Austrian psychoanalyst, who pioneered the discipline of bioenergetic analysis, a body-oriented form of psychotherapy, around 1930. It turned out to be his last coherent contribution to psychology. Over the next decade, he became increasingly irrational and restless. He moved first to Norway and then to America just before war broke out. In New York, he started having visions of "orgone," a blue particulate form of energy that he claimed was in all living things as well as the earth and sky. In 1942, he left New York and bought a property in Maine, which he named Orgonon. The land housed his residence, a laboratory and a small research community. He continued to work on isolating and accumulating orgone energy as well as developing therapeutic applications. The most iconic product of his research was his orgone accumulator, a metal-clad box that patients would sit inside to be cured of various ailments. He started to manufacture these and ship them to clients. It was about then that he caught the attention of the FDA, which opened an investigation on the basis of fraudulent medical claims.

Meanwhile, Reich's confabulations continued. In 1951, he discovered the sinister twin of orgone energy: deadly orgone radiation. It seemed that atmospheric accumulations of this radiation were responsible for the creation of deserts. But Reich, ever resourceful with his metalworking tools, had the antidote. He built an inverted form of an orgone accumulator that he called a "cloudbuster." It consisted of a double row of four 15-foot aluminum tubes attached to a base that could be rotated to point in almost any direction. The base was connected to a dozen or so grounding cables, the ends of which had to be immersed in water or soil in order for the cloudbuster to work. It looked like a cross between an antiaircraft gun and a barrage rocket launcher. There were no chemicals, projectiles or electric currents. To destroy deadly orgone radiation, he simply had to aim his cloudbuster at the sky. He also claimed the device could produce rain.

When Reich's neighbors caught wind of what he had been up to in his laboratory, they were more than credulous. In fact, during a local drought in 1953, two nearby farmers hired Reich to produce rain for their desiccated blueberry crop. He set up his cloudbuster on the morning of July 6 near their farms, and, according to an eyewitness account in the Bangor *Daily News*, it rained that evening. There was a new rainmaker in America.

Meanwhile, back at the laboratory, all was not well. At night, above Orgonon, he began to see slim, cigar-shaped alien vessels trailing deadly orgone streams. It was obviously a task for his cloudbuster, which became a weapon at the forefront of what Reich called "a full scale interplanetary battle." In 1955, he moved to Arizona where he used two cloudbusters to shoot down UFOs, or "energy alphas" as he called them. The year after, in 1956, he was arrested and ultimately jailed by the FDA. He never used his cloudbusters again, though even today his original machine sits unused and remarkably intact near the edge of a pine forest on the grounds of Orgonon.

THE REAL RAINMAKERS

While Wilhelm Reich was designing orgone accumulators in Maine, a trio of American scientists were on the verge of producing rain by tampering with clouds. Their laboratory was less than 200 miles west of him, in Schenectady, New York. The research was being run by Irving Langmuir, a Nobel laureate in chemistry, who was working under military contract to research cloud formation. He had two assistants, Bernard Vonnegut and Vincent Schaefer. Bernard's brother, Kurt, also worked at the lab writing press releases.

One of the Schenectady experiments centered on an artificial, supercooled cloud that was kept suspended in a freezer. Langmuir was trying to coax the water droplets to crystallize without having to lower the temperature of the freezer to -40°C. Schaefer was running the expermiment and had been adding various chemicals to the cloud chamber without success. On a very hot day in July 1946, the freezer motor seemed to be laboring to maintain the supercooled cloud, so Schaefer thought he'd expedite the cooling process by dropping a chunk of dry ice into the freezer. The experimental cloud instantly crystallized; he could see the light glinting from millions of sparkling crystals still suspended in the air. He tried smaller and smaller pieces, discovering that ridiculously small amounts still triggered the mass crystallization process. The dry ice initiated a sort of chain reaction where the results far exceeded the input. This was a breakthrough.

Langmuir, upon hearing about the crystallization reaction from Schaefer, immediately realized the potential: "We've got to get into the atmosphere and see if we can do things with natural clouds." On November 13, 1946, they rented an airplane, and Schaefer flew through a cold cloud hovering over the Berkshires. He dumped six pounds of dry-ice pellets over a three-mile section of the cloud as Langmuir watched with binoculars from an airport control tower several miles away. What Langmuir saw amazed him. Snow began to fall from the cloud in thick, white columns.

Kurt Vonnegut must have sent out the press release that afternoon, because the very next day the *New York Times* ran a very enthusiastic piece about the breakthrough. A few months later, in January 1947, Bernard Vonnegut discovered that silver iodide was even more effective than dry ice. Speculation in the press ran amok: custom blizzards could be delivered to ski resorts, drought-stricken farmers could irrigate their land with snow. The commercial potential was enormous, and Langmuir and his team were on the verge of becoming very wealthy. Then, in February 1947, the defense department made its move — all cloud-seeding experimentation was transferred to military jurisdiction and the Schenectady lab was shut down. The code name for the cloud-seeding initiative was Project Cirrus. If it was successful, they could weaponize the weather.

The sky was the limit in the early days of Project Cirrus or, perhaps more correctly, the stratosphere was the limit. Drunk with hubris, the defense department looked for a suitable demonstration of the power they now possessed. They needed a worthy adversary — a really big cloud. What if they could tame the mightiest storm nature could throw at them and subdue the fury of a hurricane? After all, a hurricane is simply a linked series of giant cumulonimbus clouds, and cumulonimbus clouds could easily be seeded.

THE CAPE SABLE HURRICANE

In October 1947, the perfect opportunity presented itself — the eighth hurricane of the season. After making landfall in southern Florida at Cape Sable, it sped northeastward over southern Florida, causing major flooding, and then headed out to sea. On the morning of October 13, when the hurricane was 350 miles offshore and presumed to no longer be a threat to any inhabited regions, Project Cirrus launched three aircraft carrying 180 pounds of dry ice. When they arrived at the storm, Lieutenant Commander Daniel Rex had the aircraft drop 80 pounds of dry ice on the southwest portion of the shelf cloud. He then targeted two convective towers, 60,000-foot cumulonimbus clouds, with 50 pounds each. Almost instantly, the clouds began to transform. He reported a "pronounced modification of the cloud deck seeded."

The dry-ice dump affected almost 300 square miles of the storm's cloud shield, and the airplanes headed back to their base in Tampa, Florida. According to plan, the seeding should have destabilized the convective flow of the hurricane, causing it to dissipate or at least lose strength. Instead, on the evening of the 14th, the storm pivoted and changed course, heading straight back to Georgia where it made landfall the next morning. Fortunately, it was only a category 1 hurricane when it hit. Although 1,500 buildings were damaged, there was no significant flooding. Project Cirrus was quietly canceled.

But the U.S. military was not finished with its weaponized weather mandate. Its final use of cloud seeding was during the Vietnam War: a covert offensive named Operation Popeye initiated in March 1967. For the next five years, the 54th Weather Reconnaissance Squadron regularly seeded late-season monsoon clouds over the Ho Chi Minh Trail, extending the rainy period by 30 to 45 days and making life miserable for Vietcong soldiers using the trail. The slogan for the operation was "make mud, not war." Five years after Operation Popeye was shelved in 1972, during the Environmental Modification Convention in Geneva, the U.S. signed the international treaty banning weather warfare. The treaty came into effect in 1978.

Even so, commercial cloud seeding continued throughout this period. After the failure of Project Cirrus, cloud-seeding technology became available to commercial businesses. The longest-lasting weather modification company in the United States, North American Weather Consultants, got its start in 1950 and has been seeding clouds over Utah ever since. Today it is a flourishing company, and weather scientists in Utah estimate its efforts add about 250,000 acre-feet to state rivers and reservoirs annually. North American Weather Consultants is just one of many rainmaking enterprises. Cloud seeding in the United States today is almost as routine as crop dusting. Many other countries, including Australia, Morocco, Senegal, Germany, Russia, Kuwait, United Arab Emirates, India, Indonesia, Malaysia and Thailand, have also invested in weather modification technology. But the world's biggest cloud-seeding operations are in the People's Republic of China, where rockets loaded with silver iodide are fired into clouds. During the 2008 Summer Olympics, they squeezed the rain out of clouds upwind of Beijing in order to ensure dry weather for the opening and closing ceremonies.

Of course, if there are no clouds, weather modification is out of the question. There are places in the world that haven't seen clouds, let alone rain, for decades. One of these, perhaps the driest city in the world is Arica, Chile, located at the north end of the Atacama desert. Even though it is on the Pacific coast, and next to mountains that should provide plentiful orographic rainfall, it is bone dry. During one 14-year period, from 1903 to 1917, it didn't get a drop of rain. And in the interior reaches of the Atacama are regions where rain hasn't fallen in 400 years.

This sounds like a meteorological contradiction, but it isn't. The hyperaridity of the Atacama desert is a result of the convergence of three climatic factors. First of all, it is in the middle of the South Pacific anticyclone high-pressure zone, a semipermanent region of sunny weather. Second, the cold Humboldt current upwelling offshore prevents any convective cloud formation over the ocean. And third, the region is the unfortunate victim of a reverse rainshadow effect working from east to west — the Andes block any moisture coming in from the Amazon rainforests.

Curiously, another Chilean town called Bahía Felíx, located some 2,500 miles south at the opposite end of Chile, is the world-record holder for rainy days, with an average of 325 rainy days a year. How could one country contain such extremes of precipitation? Chile is deceptively sized. Though narrow, it is long. If you reversed Chile from south to north and transposed it by latitude onto North America, then Bahía Felíx would sit in Hudson Bay and Arica would be next to Jamaica.

The uniqueness of arid Arica is exemplified when you compare it to some of the rainiest places on Earth, many of which lie on the windward side of mountains, especially those on the Pacific Ocean. Port Renfrew, for instance, on the west coast of Vancouver Island, gets 138 inches a year. But the western side of Mount Waialeale on the island of Kauai, Hawaii, in the middle of the tropical Pacific gets three times as much, about 460 inches per annum. Réunion Island in the Indian Ocean also gets massive rainfalls. In one 24-hour period, March 15–16, 1952, the town of Cilaos received 74 inches of rain. The current world-record holder for annual rainfall is a village called Cherrapunji in northeastern India: it received 1,042 inches of monsoon rain in 1860. A record that is still unbroken.

I'm not sure what the local idiomatic term for heavy rain is in Cherranpunji, but in North America, during a heavy rain, people say that it's *raining cats and dogs* or *bucketing*. Welsh downpours come in *old women and walking sticks*, and Czechs and Slovaks say it's raining *wheelbarrows*. Australians call a hard rain a *frog strangler*, and in Scotland, a torrential rain is said to be *chuckin' it doon*. Greeks say it's raining *chair legs* while in Germany a significant rain comes down in *young cobblers*.

NERO'S RAIN PALACE

Leave it to the Romans to turn the concept of shelter inside out or, should I say, outside in. The Julio-Claudian dynasty, which began with Augustus and ended with Nero, was famous for its excesses. Nero embodied the pinnacle of imperial immoderation. He fancied himself a superlative singer, often competing in musical competitions as far afield as Greece. Nero also held private concerts for the nobility in Rome. According to the Roman historian Suetonius in his tell-all *The Lives of the Caesars*, the only way to escape the tedium of Nero's lengthy solo performances was to feign death or jump from second-story lavatory windows.

Nero, like many other Roman emperors, took an interest in architecture. He loved to pore over scale models with his personal architect, and he was particularly keen on an ambitious design for a palace he'd spent years sketching out. But vacant real estate was in short supply in Rome, and there was nowhere to erect such a large building. Until 64 CE, that is.

Since the fiddle had yet to be invented, it is impossible that Nero played it while Rome burned. He did, however, take the opportunity presented by the freshly available land in the center of Rome to construct his architectural fantasy. He named his palace Domus Aurea, and it was centered on a grand hall, the *coenatio*, where Nero appeared during private rituals celebrating his deification. Above him, inside the domed ceiling of the *coenatio*, he commissioned two of Rome's foremost architects, Severus and Celer, to build a series of nested semispherical, rotating domes that simulated the movement of the planets in the night sky. As Suetonius wrote, "The main chamber was round and revolved continually day and night, as does the world." Nero also had his engineers incorporate an ingenious network of metal tubing into the mobile domes so that, on his command, rain would fall from his artificial sky.

Seneca, who was Nero's preceptor, described these mechanisms: "A technician invented a system by which saffron-colored water poured down from a great height and also succeeded in assembling the panels of this chamber's ceiling in such a way that it changed at will." Unhappily for Nero, he was despised by the Flavian dynasty that supplanted him after his suicide, and his works were destroyed in a retributive policy of *damnatio memoriae.* His magnificent *coenatio* along with the rest of the Domus Aurea was reduced to rubble.

FREEZING RAIN

Ice storms are destructive, delicate creatures, requiring perfect conditions before they can deploy their dangerous alchemy. First of all, the surface temperature has to hold at between -1°C and 1°C for the duration of the ice storm, which usually lasts 12 to 20 hours. High above the ground, the clouds have to be stratified with exactly the right thermoclines — alternating layers of warm and cold air — to ensure that the rain becomes supercooled just as it leaves the cloud. Finally, freezing rain is enhanced if the temperature of things on the surface — trees, roofs, cars, buildings — has been maintained at below-freezing temperatures for at least 24 hours prior to the storm. This subzero reservoir makes the supercooled rain freeze on contact, like transparent glue falling from the sky. After that it's just a question of accumulation.

When a tree branch or a power-transmission wire is coated with ice, its surface as well as its mass is multiplied. This means that the load factor increases exponentially with every centimeter of ice that adheres, doubling and then quadrupling the pull of Earth's gravity, an inexorable, devastating force. Ice storms should really be called gravity storms.

For trees, the destruction from a moderate ice storm is the same as that produced by an F3 tornado, with every tree sustaining some damage and many trees losing main branches. The difference is that the path of an F3 tornado is normally less than a hundred feet across and a mile or two long at the most. By contrast, the great ice storm of January 1998, which struck eastern North America, affected thousands of square miles. During that storm, four inches of freezing rain collected on everything in the system's path; hydro towers crumpled and whole forests in eastern Ontario, southern Quebec and upper New York state splintered into kindling. More than three million people were left without power for days, and thousands were without power for weeks afterward. The overall damage amounted to $1.4 billion. I missed that one, but I've been through several others.

Ice storms can seem innocuous at first. The rain — usually light, always steady — often sets up the evening before the storm. That was true of the most recent ice storm I experienced in Toronto a few years ago. Just after sunset, the ice storm began in a spectral quietude, a hush that arrived with the misty precipitation. It was hard to tell at first if the rain was indeed freezing, but then, later that night, I opened a window and heard the unmistakable sound of ice creaking as a mild breeze swayed the branches.

I wasn't sure how serious the storm was until the next morning. Around 7:30, I was woken up by a loud cracking noise followed by a glassy crash, as if a car had driven through a plate glass window just a few houses away. Minutes later, the same noise. I looked out the window. The trees were collapsing. I put on my bathrobe and went outside. Some of my neighbors were already sprinkling salt on their steps and gawking at the devastation. "At least we still have power," said Evelyn from across the street. It was a cataclysmic scene: huge branches down and a severed power wire sparking on the ice at the end of the block. One of the bigger branches had crushed the rear of a parked car; another lay across the street, completely blocking it. Everything was coated with ice. It was surreal, apocalyptic. When I went back, I found that the power was out.

I started a fire and carried in several armfuls of what little firewood I still had in the garage. My fireplace would be my only source of heat. If the blackout went on for too long and the temperature dropped, my pipes might freeze along with my indoor plants. Ice storms are the reason many of my neighbors had invested in gas fireplaces — during a blackout, the gas still flows.

Fortunately, the power came back on a few hours later, although my house had started to chill despite my cheery fire. That night, I heard more crashing branches, but the next morning revealed an entirely different scene. A cold front had moved in overnight, the temperature had plunged and skies were blue. My backyard had transformed into an dazzling tableau of cut crystal. The icy branches and phone wires sparkled with rainbow prisms in the sunlight. (I noticed that the hues of the ice spectrum were confined to two colors — yellow and purple — a result, I'm guessing, of the limited refractive properties of frozen rain.) By noon, the ice began to fall from the sun-warmed branches, clattering down on the sidewalks and icy yards. By sunset, the neighborhood was littered with drifts of split ice tubes: perfect molds of the branches and wires they had once clung to.

5

THE SECRET LIFE OF STORMS

"I saw the lightning's gleaming rod
Reach forth and write upon the sky
The awful autograph of God."
JOAQUIN MILLER

Our storm-prone globe would be a hard sell for interstellar real estate agents. Imagine a pair of newly married aliens, eager to buy a house and start a family on our lovely blue planet. Everything looks good as their agent goes over the selling points: plenty of water, equitable mean global temperature of 14°C, lots of fascinating life-forms and landscapes — a virtual paradise. The politics and economics of the earthlings hold little interest for the aliens, but their attention is piqued when the agent begins to describe planetary atmospheric conditions. "Statistically speaking," he says, "the atmosphere is quite hospitable." Then he adds that there are "very *occasionally*" (as he is quick to qualify) "weather disturbances referred to as *storms*. Nothing really to worry about, though some of them are quite a nuisance."

"Now I *am* worried," says the female alien. "What kind of nuisance?"

The agent looks uncomfortable. He elaborates, "Well, most storms involve an electrical phenomenon called lightning, large bursts of energy that conduct a differential charge from the cloud to high points directly under the storm." He has her full attention.

"How large?" she asks.

"In the order of several hundred million volts," he replies.

Eventually they wheedle the ugly truth out of him, not just about lightning but also hurricanes, tornadoes, hail, typhoons and monsoon rains. The planet is riddled with storms that can strike anywhere at random, on land or sea. In fact, at any given moment, there are about 2,000 active thunderstorms worldwide, and the human population has no control over them whatsoever. The deal-killer for our alien couple is learning that storms kill hundreds, sometimes thousands of earthlings every year.

"What about Mars?" the male alien asks. "We passed it on the way in."

"Well," the agent replies, "I must caution you that it does have dust storms, a lot of sand and wind, but no lightning and nothing like hurricanes or tornadoes. Just a few oversize dust devils. It's cold, dry and has lovely, pastel sunsets. You might like it." Seeing their polite hesitation, he adds, "At any rate, we have another office quite nearby actually, in the Alpha Centauri system, just a little over four light years from here."

As permanent residents of Earth, we don't have the option of moving elsewhere. For better or worse, we're stuck here. And it could be worse. Take Jupiter, where millions of electrical storms rage daily. In fact, Jupiter's Great Red Spot is actually a stupendous, stationary hurricane with a 19,000-mile-wide eye that has raged for at least 300 years. Saturn is little better. When *Voyager* passed by in 1981, it picked up static from a giant equatorial thunderstorm 37,000 miles wide, with wind speeds that clocked in at over 932 miles per hour. We've got it easy.

Even so, earthly storms can pack a punch. A typical thunderstorm produces several hundred megawatts of electrical power, enough to supply all of the U.S. for 20 minutes — or by a more violent yardstick — it's equivalent to the energy released by an atomic bomb. So when a storm looms on the horizon, all we can do is batten down the hatches or, in the case of tornadoes, duck into a storm cellar if we have one. But how do storms begin? Why are they so violent? The answer has a lot to do with heat. Tornadoes, hurricanes, typhoons and most thunderstorms need warm weather to spawn, as well as moisture, and a lot of it. Storms are clouds showing their muscle.

A summer thunderstorm starts life as a cumulus cloud cruising over fields and rivers and lakes. It's a case of wandering lonely as a cloud, certainly, but a cloud with megalomanical tendencies. The fledgling storm feeds on thermal updrafts, sucking up moisture like a vacuum cleaner, and, as it drifts, it inflates until it reaches the next stage of its evolution, transforming into a cumulus congestus. This is a sort of adolescent cumulonimbus that can reach heights of 6,500 feet. Now the first drops of rain begin to fall.

At this point, two things happen. The rising humid air that initially puffed up the cloud now accelerates into an industrial-scale updraft, while rainy downdrafts begin shunting cooler air downward at the same time. All this commotion creates friction, particularly at the atomic particle level, and static electricity begins to build up charges in the cloud. Our cumulus congestus has become a full-fledged cumulonimbus and now has electrical potential — a positive charge in the cold heights of the cloud and a negative charge in the warmer regions closer to the Earth. Zeus's arsenal.

These charges build in strength until a glowing stream of electrons irresistibly bridges the gap. Zeus hurls a bolt.

But air is not a good conductor, even with billions of volts and hundreds of thousands of amperes at a bolt's command. The atmosphere might as well be a thick rubber sheet. Lightning needs help, and Joseph Dwyer, a physicist from the University of New Hampshire, thinks it might come from cosmic rays — high-energy particles moving at the speed of light out of the cores of exploding stars. Our planet is continuously bombarded by millions of cosmic rays; as they streak through the atmosphere, they leave a trail of electrically charged particles. These brief channels are what lightning rides to the ground, and sometimes into it. Lightning often penetrates yards into sandy soil, fusing the mineral particles together into solid, wiggily tubes called fulgurites, or lightning stones.

LIGHTNING

"Thunder is good, thunder is impressive;
but it is lightning that does the work."
MARK TWAIN

My childhood home in southwestern Ontario was on the crest of a ravine that opened westward over an expanse of oxbow lakes. It was a perfect vantage for watching summer storms. My mother, born in Medicine Hat, Alberta, had prairie experience with tornado weather, and on hot, humid days with a strong southwest wind, she would tell my brothers and me to keep an eye on the western horizon starting around four in the afternoon. A severe storm could blow up quickly and it was best to have a safe place handy to wait it out. I think that, like me, my mother privately enjoyed thunderstorms.

If a tornado actually touched down, my mother told us the most protected place in any building was the southwest corner, which made sense — tornadoes usually track from southwest to northeast. Despite her admonitions, when the weather was right I prayed for storms.

There was a delirious excitement that came with the approach of a cumulonimbus cloud, particularly if it was a "line storm." These storms were always big and had a characteristic shape. Meteorologists now call them squall lines. There's a wonderful oil painting of one by the American painter John Steuart Curry, titled simply *The Line Storm*. It depicts a summer afternoon in the American Midwest, grain fields, tree-lined country roads, barns and a windmill, above and behind which looms the black, bowl-shaped front of a line storm rising above the horizon. The storm occupies the upper half of the painting. In the foreground is a horse-drawn hay wagon racing to find shelter. Curry's painting captures the electric mood of apprehension at the imminent approach of unknown violence.

During the summer, we'd get our share of regular thunderstorms, but we were also hit by line storms, like the one in Curry's painting. These would strike, just as Mother predicted, around four or five in the afternoon. The hot wind would die suddenly, and then the curved line of the storm would begin rising above the western horizon like the leading edge of some inconceivably huge flying saucer. Underneath the projecting upper rim was an ominous blue-black zone that widened as the storm approached. You could see lightning flickering in there, and as the storm grew nearer still, you'd hear the first rumbles of thunder.

I'd run to my vantage on the edge of the ravine where I loved to watch the edge of the storm slide overhead, the point of no return. On the street behind me, adults rolled up their car windows and snatched down any laundry still out on clotheslines while the thunder got louder and the flicker of lightning started to cast sudden, alarming shadows. This was drama. Then it got darker still, and the streetlights came on, and the sky took on an eerie purple or greenish cast. Invariably, there would be a close lightning strike where you could see the whole bolt, followed immediately by a loud crack of thunder. About a mile distant, in the darkness under the storm, the rain curtain approached just behind the lashing treetops. The squall line was coming. I could count down from that point, about 20 seconds, before the wind hit and leaves and small branches started flying. But I stood my ground at the top of the ravine. It wasn't until the first big drops started to fall that I retreated to my house in time to avoid being completely drenched.

Viewing the rest of the storm from the relatively safe vantage of my parents' glass-walled sun porch, I'd watch the street turn into a river. Lightning was now constant, and the windows of the porch rattled with the thunder. Neighbors caught in the downpour splashed along with their jackets pulled over their heads. Soaked to the bone, as we used to say. I thought of monsoons in southern India, the tropical foliage, the wet arms and legs. Then a terrific flash-bang, very close. Did that hit a house in the neighborhood? I would try to figure out how far away the center of the storm was by using a technique my father once taught me when we were camping.

Temperatures at the core of a lightning bolt can be as high as 30,000°C, or six times hotter than the sun's surface, instantly vaporizing the narrow tunnel of air it travels down. The surrounding air is pushed outward by the heat, and the shockwave travels at the speed of sound until it reaches our ears as thunder. By counting the seconds between the flash and the thunder, every five seconds being equivalent to a mile, you can determine how far away the lightning strike is. A direct, vertical cloud-to-ground strike (or a "CG" as storm chasers like to call it) will produce a single loud clap of thunder, whereas branched lightning, sometimes completely confined to clouds, creates long, rolling peals of thunder. With a stopwatch and a little experience, you can map out the height and length of branched lightning bolts. But this is academic if you are camping and prone to a little anxiety about storms. The shortening seconds between flash and thunder only increase your panic. Who knows where the next bolt will strike as the center of the storm gets closer?

Lightning is dangerous. On average, lightning kills 16 people a year in Canada and 95 in the United States. It doesn't seem like a lot compared to automobile accidents or gun violence, but it is sometimes a gruesome way to go. A friend of mine attending an equestrian event on a cloudy summer afternoon was close to another rider when she was struck by a rogue bolt of lightning. The horse survived but the rider didn't. What my friend couldn't get out of her mind, for weeks afterward, was the smell. "Like a roast," she said. And it wasn't even raining.

If you are caught out in the open by an electrical storm, never take shelter under a single tree. A thicket of short trees of uniform height theoretically offers better protection, but lacking that it's best to crouch with your feet and knees together in the open. Some guides say it doesn't matter if you stand or crouch, but I say crouch because you have a lower profile that way. I'll tell you why.

Years ago, I was in a log cabin in northern Ontario that was struck by lightning during a pop-up storm that had come in fast over the lake. The concussion of thunder (which must have originated within the cabin) was like an explosion. It deafened me. Then I remember that the air was suddenly hazy, my ears were ringing and one of the logs in the wall had exploded. I waited out the rest of the storm outside the cabin. It was irrational, but I was convinced that lightning would strike there again.

The cabin survived, but, upon inspecting it half an hour later, it was remarkable to see the paths of the various tendrils of lightning. Besides scorching and stripping a spiral of wood from a log in the wall, it had splintered floorboards and even run through the metal wires on a dartboard, jumping from there to split a mercury thermometer sitting in a jar beneath, a yard from where I was standing. It was sobering. I had been lucky.

My childhood intoxication with electrical storms changed that day. Now, I really don't think you want to be any closer to lightning than you have to be, so I say crouch if you're caught in a field. Why with knees and feet pressed together you might ask? Because if your feet are even slightly apart, it will tempt the lightning to pass through your body. Flesh and blood are a much better conductors than grass and soil. The interiors of cars are safe, because the metal body conducts the strike around the occupants and to the ground. Large enclosed spaces are still the safest place to be during electrical storms. In the Middle Ages, it was thought that a Yule log kept in the fireplace during the summer warded off lightning strikes, but these days just staying off a landline telephone during a storm will suffice. Curiously enough airplanes are also generally safe, even if they sustain a direct strike.

"All night sheet lightning quaked sourceless to the west
beyond the midnight thunderheads, making a bluish day
of the distant desert, the mountains on the sudden skyline
stark and black and livid like a land of some other order
out there whose true geology was not stone but fear."

CORMAC MCCARTHY

Earth plays host to about 44,000 thunderstorms daily, which means that it is being struck by lightning around 100 times a second. Lightning comes in a plethora of shapes and sizes — heat lightning, spider lightning and even hot and cold lightning. I liked the idea of cold lightning when I first heard about it, imagining a cool neon bolt with thunder and no heat, but that isn't anywhere close to what the term means. The difference between hot and cold lightning turns out be the number of return strokes, and their duration, in a single lightning bolt. It all becomes visible when lightning is slowed to a crawl.

Filmed in extreme slow-motion, lightning doesn't even look

like lightning, and there is way more going on than you'd think. The ghostly, sparking tendrils of the main leader snake randomly in all directions as it approaches the ground, exploring the potential of the space around the bolt. The tendrils that don't connect fizzle out while the one that does touch the ground suddenly turns white hot as the entire billion-volt charge zooms down the corridor opened up by this single successful filament. Then a brief pause of darkness and suddenly again the same shape is illuminated as a return charge races up to the cloud. If this happens just once, it's cold lightning. But if several strokes of longer duration follow the same path upward and downward, it's hot.

Return strokes were first photographed in 1902 when Sir Charles Vernon Boys used a special camera he'd invented that used revolving film to capture lightning bolts in a time "smear." When the moving film is exposed to the light from a single lightning strike, it spreads it out in time so that the resulting photograph of the various return strokes looks like after-images of the same bolt repeated side by side. More return strokes means longer duration and more potential for whatever the lightning hits, be it dry grass or wood, to catch fire. That's why it's called hot lightning.

Hot lightning is not to be confused with heat lightning, the silent lightning that accompanies storms during heat waves. Years ago, during an early summer heat wave, I witnessed several nights of magical heat lightning in the skies over my hometown. For two or three evenings one week, the night sky would fill with low, rainless storm clouds and then, for hours on end, there would be a spectacular light show of spider lightning (lightning with many branches and channels that sometimes recombine) so prolific it looked like electric lacework stitching the clouds. What made it eerie was that the entire spectacle was completely silent.

It would be remiss of me not to mention that what I witnessed in my hometown is officially impossible. The meteorological consensus on heat lightning is that it's merely lightning that is too distant to hear the thunder. But much of the heat lightning I saw was directly overhead, less than a mile distant.

RIDERS ON THE STORM:
SPRITES, BLUE JETS, ELVES, PIXIES, GNOMES AND TROLLS

But the story of lightning gets stranger and stranger. There is another form of lightning, one that doesn't jump from cloud to cloud or cloud to ground. Instead it reaches straight up to the heavens. High above the fury of violent thunderstorms, in the calm, sparse air of the stratosphere and beyond, ghost riders flicker and dance like vanishing hallucinations.

Some of these apparitions, pale as faint red neon signs and shaped like jellyfish, appear fleetingly, as if momentarily lit up by a stray flash. Others look like blue rays speeding upward like rockets from the top of an anvil cloud. A third, more massive type illuminates the lower ionosphere in a pale 124-mile-wide pink donut. All are ephemeral.

These electrical specters were discovered accidentally a little over two decades ago by scientists from the University of Minnesota. On July 6, 1989, they were testing low-light video cameras at night and by chance caught an image of a peculiar glow above a midwestern storm. Later analysis proved the image to be no optical flaw; something extraordinary had appeared in the sky over the storm clouds.

These aerial lights are known, rather poetically, as sprites and blue jets. (Yes, scientists can be whimsical too.) As it turns out, there had been previous eyewitness accounts of these strange coruscations. Sprites have been dancing above storms for millions of years, so it stands to reason that a few humans had already seen them. It's just that these accounts were so sporadic no one thought to investigate them.

The first written record of what was most likely a blue jet was published in 1730 by Johann Georg Estor, a German legal theorist, in the second volume of his *Neue Kleine Schriften.* He described climbing a mountain in the Vogelsberg up through a thunderstorm. When he emerged from the top of the storm cloud and looked down on it, he saw something he'd never seen before — a streak of lightning shooting straight up into the clear sky above the storm.

Since Estor's sighting, there have been handful of accounts by various observers, of both blue jets and sprites, but it took the University of Minnesota video to get the ball rolling. Over the past two decades, many more of these enigmatic lights have been observed, and several distinct types have emerged from the research. The first, witnessed by the night camera in 1989, was named a sprite by the two leading experts on this form of lightning, Davis Sentman and Eugene Wescott of the University of Alaska at Fairbanks. Their folkloric nomenclature stuck, and, as a result, the names for upper atmosphere lightning manifestations sound more likely to appear on a page in *The Lord of the Rings* than a scientific treatise.

Even their rival, the meteorologist Walter A. Lyons, fell in line. His scientific label for the pink donut lightning he discovered was emissions of light and very low frequency perturbations from electromagnetic pulse sources, or ELVLFPFEPS, but he shortened it to ELVES. Since then, other mythological creatures have been added to the family — pixies and gnomes (briefly glowing balls of light on the upper edges of storm clouds) and TROLLS (transient red optical luminous lineaments), the red tendrils that hang down from sprites.

All of these ethereal forms of lightning are completely unlike the mighty bolts that singe the clouds beneath them. They are radiant plasma, more like the glow within a fluorescent light. But they are colossal. Hovering between 25 to 56 miles above the Earth, sprites are often 31 miles tall, and some ELVES are more than 245 miles in diameter. What they lack in voltage, they make up for in pure scale, dwarfing the storm clouds that birthed them. Sprites are not completely without sting, however. The sudden electrical failure of an upper atmosphere NASA probe on June 6, 1989, has been retroactively linked to damage from a sprite. Come to think of it, was William Rankin's inexplicable engine failure over the thunderstorm in North Carolina due to a blue jet overloading all the electrical cables in his aircraft? We'll never know.

GREAT BALLS OF FIRE

Just after midnight, on March 19, 1963, Eastern Airlines flight 539 was en route from New York City to Washington, D.C., when it encountered a turbulent thunderstorm. The passengers barely had time to fasten their seatbelts before a sudden burst of light and a terrific thunderclap shook the plane. Moments later, a glowing sphere, about the size of a volleyball, floated out of the pilot's cabin and drifted down the aisle to the back of the airplane where it disappeared.

This would have been tabloid fodder except that one of the passengers was an English physicist and author. His detailed account of the incident in the prestigious journal *Nature* gave new credibility to a phenomena that had previously inhabited a mythic realm. Ball lightning turns out to be one of the most scientifically contentious of all exotic electrical atmospheric phenomena, even with hundreds of eyewitness accounts. In almost all of these reports, glowing spheres float above the ground and drift along at a walking pace until they disappear or, in some instances, explode.

The British occultist Aleister Crowley was one such eyewitness. He had a close encounter with ball lightning one summer in 1916 while staying at a cottage on Lake Pasquaney in New Hampshire. Though not a physicist himself, his lucid account of the encounter would do credit to any scientist. He'd been caught outside by a sudden downpour and had rushed indoors to change into dry clothes as the storm raged outside.

To put on my stockings, I sat down in a chair close to the brickwork of the chimney stack. As I bent down, I noticed, with what I can only describe as calm amazement, that a dazzling globe of electric fire, apparently between six and twelve inches in diameter, was stationary about six inches below and to the right of my right knee. As I looked at it, it exploded with a sharp report quite impossible to confuse with the continuous turmoil of lightning, thunder and hail . . . I felt a very slight shock in the middle of my right hand, which was closer to the globe than any other part of my body.

A diameter of eight inches seems to be the norm for lightning balls, but that's about their only regularity. They don't appear to obey the normal laws of electrical physics. Like ghosts, they can pass through conductors (screens or metal doors) and non-conductors (glass windows, brick walls) without leaving any trace, though in 1944, when ball lightning passed through a window in Uppsala, Sweden, it melted a perfectly circular golf ball–sized hole in the glass. Sometimes, the balls float above the ground, meandering through the air, and at other times they seem to roll along surfaces. Witnesses say they emit a fainting hissing noise, "like a flaring match," recounted one man. They also come in a range of colors — violet, white and pale blue, though mostly varying from red to orange to yellow. These kinds of inconsistencies are what make lightning balls so maddening for scientists. They defy categorization. How can the electrically charged plasma hold its perfectly spherical shape for so long? What exactly is happening inside them?

WHISTLERS

If you take an AM radio and tune it between stations you'll hear static — little bursts of crackling noises, some faint, some louder.

This is the sound of lightning from distant storms arriving in the form of the radio-frequency pulses, or radio signatures. They're called sferics, and the faintest are thousands of miles away. But if you have a tuner that can receive VLF (very low frequency) radio signals, and you tune it between stations (there aren't a lot of them), you'll likely hear eerie descending whistles like the sound of bombs falling. These are sferics that have leaked out of the lower thermosphere and traveled along the magnetic field lines at the edge of the exosphere, 6,000 miles above Earth's surface and back again. On their trip through the magnetized plasma in the exosphere, the sferics get dispersed, stretched out and arrive back at Earth in a series of descending tones instead of a single static crackle. Sometimes the same whistler will bounce back and forth, caught in the magnetosphere, the descending tones getting longer and fainter until the whistler dissipates entirely.

HAIL

Upper atmosphere storm phenomena such as whistlers and pixies are marvelous, but they are irrelevant to us who must endure thunderstorms from beneath, here on the surface of the planet. We bear the brunt of nature's fury, and as if wind, lightning and flooding weren't enough, there are other dangers. The powerful updrafts within storm clouds that our pilot William Rankin experienced firsthand are also hailstone factories.

Supercooled water combines with ice crystals in the cold heights of a storm cloud to form ice pellets. These ride downdrafts until they partially melt, but then updrafts shoot them back up into the cold zone where they freeze again. They then fall once more before hitting another updraft and acquiring another coat of freezing rain. This keeps happening until they reach a size that can no longer be held aloft by an updraft. Repeated cycling up and down through the storm cloud, where the hailstone freezes and melts, builds alternating layers of clear and milky-tinted ice. If you slice a hailstone in half, it has concentric growth rings, like a tree trunk.

Updrafts can reach speeds of 100 miles per hour in a big storm, and golf ball–sized hailstones are not unheard of, smashing greenhouses and car windshields. A supercell updraft can be even more powerful. Hail can get dangerously large. On June 22, 2003, a two-pound hailstone, seven inches across, fell in Aurora, Nebraska. But the world record for the heaviest documented hailstone goes to a lump of ice that fell in Bangladesh in 1986 during a hailstorm that took the lives of 92 people. It was a shade more than 2.5 pounds. A single hailstone that size would be a hazard, but imagine a few thousand of these melon-sized chunks tumbling out of the sky. An umbrella would be useless, and you wouldn't be safe even indoors. With a terminal velocity somewhere in the range of 105 miles per hour, they would easily smash through most wooden roofs. It would be like having a thousand major league pitchers hurling bricks at you.

Even mothball-sized hail is not a good sign. Hail is the trademark of a truly violent storm; in a supercell thunderstorm, the hailstones fall thickest just before the leading edge of the mesocyclone, where a tornado is most likely to be. There's a lot going on inside a cumulonimbus cloud, and ultimately hail turns out to be the least of it.

"IT'S A TWISTER, IT'S A TWISTER."

FARMHAND ZEKE IN *THE WIZARD OF OZ*

A funnel cloud is a surreal mixture of the apocalyptic, the beautiful and the demonic. Its spooky, serpentine shape panics some and mesmerizes others. Louder than a train or a jet engine, the roar of a tornado drowns out thunder. At night, they are even more ominous, sometimes glowing internally with continuous lightning like pillars of fire. What could be more cataclysmic, more threatening? Tornadoes are menacing, divine, capricious, unpredictable and wrathful.

A world apart from Kansas is Fukuoka Prefecture in southern Japan, where, in 1920, the future expert on tornadoes Tetsuya Fujita was born. Fujita was impressive. The college he attended hired him as professor of physics as soon as he graduated in 1943. He was also lucky. Two years after he began his assistant professorship at the Fukuoka Prefecture college, on August 9, 1945, an American bomber carrying the world's first plutonium bomb had to divert to its secondary target, Nagasaki, because of bad weather over the primary one — the city of Kokura in the center of Fukuoka Prefecture.

He was one of those exceptional people for whom opportunity seemed to leap out from the heart of disasters. A few weeks after the bombing of Nagasaki and Hiroshima, the Japanese defense authorities enlisted Fujita to study the blast damage at the two cities. No one in Japan had any idea what type of weapon had caused so much destruction, and all sorts of conflicting theories abounded, including magnesium flares (to account for the brightness) and multiple blasts. Fujita made an extensive photographic record of the devastation and studied the images. From the pattern of nuclear flash burns on a bamboo vase, he proved that there had only been one bomb. He also analyzed the "starburst" pattern created by flattened trees and buildings surrounding ground zero and was able to estimate not just the size and power of the blast but also how high above the ground the weapon had been when it was detonated. (According to his estimates, the power of the bomb was stupendous, almost impossible for a single detonation. His calculations reinforced the Japanese decision to surrender.) The research Fujita conducted in Hiroshima and Nagasaki, a sort of reverse engineering of disaster, was vital, if heartbreaking, training for what ultimately became his life's vocation.

After the war, Fujita studied meteorology at Tokyo University. He was fascinated by storms, perhaps as a consequence of or metaphor for the apocalypse he had narrowly escaped, and he devised radical theories about the structure and behavior of violent storms. His conjectures were advanced for the time. He read work by the American meteorologist and storm researcher Dr. Horace Byers of the University of Chicago and realized he had a kindred spirit. Fujita sent his research papers to Byers, and the two began a lively correspondence.

Then Fujita's fluky good fortune struck again. Tornadoes are as rare as hen's teeth in Japan, but one hit Kyushu on September 28, 1948, and Fujita was able to visit the debris zone almost immediately. What he discovered there was his destiny. All his training in engineering, physics and meteorology converged. He knew the load factors, the stress points, the mass and resistance of most man-made structures, so that he could deduce from a glance the scale of the forces that had acted to destroy the buildings and, from this, Fujita could reconstruct the tornado's vortex.

As soon as he was awarded his doctorate by Tokyo University in 1953 (his thesis was on typhoons), Byers invited him to work as a visiting research associate at the University of Chicago. He ended up staying and by 1955 had tenure in the meteorology department. The university sweetened the deal by purchasing a substantial brick house for Fujita and his wife, Sumiko, right beside the campus. Now Fujita could roll up his sleeves and work on mesometeorology: the study of middle-sized atmospheric phenomena such as storm cells and cyclones. But all this was prelude to his ultimate obsession: tornadoes. He became an armchair storm chaser, an academic sleuth of the tornado and of all the transient, yet indelible, traces of this capricious and terrible whirlwind.

He poured over aerial photographs of tornado tracks, photographs of wreckage, impact pits on the earth (where solid objects had hit and then bounced back up into the air), broken buildings and tree branches, and he eventually produced detailed, hand-drawn tornado paths annotated with copious notes and numbers.

He realized, after analyzing their paths, that single tornadoes almost never follow a storm for hours, but instead "families" of tornadoes form, dissipate and reform along the storm's path.

Fujita's meteorological triumph came during the massive tornado outbreak of Palm Sunday, April 11–12, 1965. Forty-seven tornadoes raged across the U.S. Midwest, killing hundreds and injuring thousands. Fujita spent months examining the aerial and ground photographs to compile his usual, meticulous damage paths. But now he saw something deeper, something more complex about these paths. There were swirls within swirls. He realized that a large, single tornado is made up of multiple vortices, smaller tornadoes that are bundled together inside the larger one. Years later, he used films of this outbreak and others to estimate the wind speeds of individual tornadoes. With these figures, and with help from Sumiko, he devised his most famous contribution to meteorology, the Fujita scale, which finally calibrated the strength of a tornado.

The original six-point Fujita scale runs from 0 to 5, where an F0 is a tornado with winds of 40 to 70 miles per hour and an F5 with winds of 261 to 319 miles per hour. An F0 might blow a few leaves off a tree and rattle a window or two, whereas an F5 leaves only stumps and empty basements. Armed with this scale and a completely fine-tuned system of debris assessment, he was ready for the Super Outbreak of April 3–4, 1974, when 148 tornadoes ravaged 13 midwestern states as well as striking southwestern Ontario.

Here, alongside the complex patterns caused by tornadoes, he discovered something new — starburst patterns of fallen trees that could only have been caused by 150-mile-per-hour downdrafts. Fujita wrote later, "If something comes down from the sky and hits the ground it will spread out . . . it will produce the same kind of outburst effect that was in the back of my mind from 1945 to 1974." He called these "downbursts" and "microbursts" and declared that there had actually been 144 tornadoes and four microbursts. (More recently, "plough winds" have been added to the arsenal of tree-snapping storm winds.)

The pilot William Rankin directly experienced the cloud machinery that produces these downbursts during his 1959 freefall into a storm. It's unlikely that he would have lived to tell his tale if he'd had the misfortune of being caught in a downdraft that ended as a downburst. He would have been dashed to earth, chute and all, at more than 100 miles per hour.

Most meteorologists received Fujita's downburst theory with skepticism, but he pressed on. It wasn't until he analyzed an aeronautical cold case that his theory gained national attention. The case in question was a famous air disaster, the 1975 crash of Eastern Airlines Flight 66 at Kennedy Airport in New York. Going back over the records of weather conditions during the disaster, Fujita realized that a downburst had hit the airplane during its approach. He then went over dozens of previous airport accident reports and combined them with weather conditions at the time. The data confirmed his theory — more than 500 air-related deaths in the previous three decades had been caused by microbursts.

Still, it took 14 more years before Fujita's data was so overwhelmingly conclusive that commercial airports installed Doppler radars to detect downbursts and reroute flights. Countless lives have been saved since. In 1991, the Japanese government recognized his work, awarding him the Gold and Silver Star, one of the highest honors of the nation.

BIRTH OF A TORNADO

Tornadoes can strike almost anywhere, from downtown London, England to mainland Greece. The earliest recorded tornado struck Rosdalla, Ireland, on April 30, 1054, and the deadliest struck two adjacent towns in Bangladesh, Saturia and Manikgank Sadar, on April 4, 1989, killing 1,300, injuring 12,000 and leaving a further

80,000 homeless. The swath of destruction leveled every structure in an area of two square miles.

Contrary to popular belief, tornadoes can strike near rivers, lakes and even mountains. In Yellowstone National Park, a tornado cleared a swath up and down the slopes of a 10,000-foot mountain.

Tornado Alley — a corridor that stretches from central Texas north into Oklahoma, Kansas and Nebraska, then eastward into central Illinois and Indiana — is the world capital of tornadoes. Here is where the majority of the 800 tornadoes that strike the U.S. annually touch down. As elsewhere in the Midwest, tornadoes occur when cold, dry air from Canada collides with warm, humid air from the Gulf of Mexico. That's a prerequisite for a lot of summer storms, but the essential element, the one that gives the twister its twist, begins hours before, under a sunny sky.

The instigator of the tornado is wind shear — winds at different altitudes moving in different directions. On a clear, humid day, these opposing layers of wind are invisible. Long before the first storm clouds begin to take shape, wind shear creates a cylinder of air, hundreds of feet high and miles long that rolls along the ground like an invisible steamroller. This is a mild presentiment of what is to come, a ghost tornado, momentarily riffling your hair on a summer afternoon.

This cylindrical, phantom tornado barrels along over fields and highways until it encounters a strong updraft that lifts one section of the spinning tube upward into the sky above the dew point. As the rotating tube rises into cooler air, a cumulus cloud begins to form around it — first as a wisp (a cumulus fractus), then as a puffy cumulus humilis that quickly inflates into a cumulus mediocris and then into an even larger cumulus congestus. In less than an hour, the cloud has become a giant cumulonimbus. Meanwhile, the rolling cylinder has been drawn up into the cloud's core in a vertical, bent loop. As the cloud continues to rise, one side of the rotating loop disappears and the remaining piece dominates. The entire central portion of the storm cloud begins to rotate in the same direction as the dominant loop fragment. Half an hour later, the storm is a giant, supercell cumulonimbus incus with a mesocyclone at its heart.

If you are watching a supercell approach from the ground, the most striking feature is the wall cloud: a two- to three-mile-wide disc-shaped cloud that projects beneath the rest of the storm. This is the visible part of the mesocyclone. It sits at the core of the storm and extends up for thousands of feet. The funnel cloud, if it forms, will drop down from the edge of this wall cloud.

Tornadoes almost always travel from the southwest to the northeast, moving at just above the average urban speed limit — 35 miles per hour — but they have been known to move at twice that speed and as slowly as five miles per hour. They have been seen traveling in zigzag, circular or looping patterns. And then there are some that remain stationary. In South Dakota, a tornado once hung over a single field for 45 minutes. In a car, you can generally avoid them, although even the most seasoned storm chasers have been trapped while driving and then paid with their lives. And yet you can see why people chase storms or don't bother to take shelter. What could be more thrilling than feeling the earth rumble beneath your feet as an F5 passes within less than a mile?

On the afternoon of June 22, 1928, Will Keller, a Kansas farmer, saw a tornado bearing down on his home. He hustled his family into their storm cellar, but, as the tornado approached, he paused at the top of the stairs with the door still open. I can imagine him, transfixed by the immensity of the oncoming spectacle while his family begged him to shut the door and come down. But curiosity got the better of him.

> *As I paused to look I saw that the lower end which had been sweeping the ground was beginning to rise. I knew what that meant, so I kept my position. I knew*

that I was comparatively safe and I knew that if the tornado dipped again I could drop down and close the door before any harm could be done.

Steadily the tornado came on, the end gradually rising above the ground. I could have stood there only a few seconds but so impressed was I with what was going on that it seemed like a long time. At last the great shaggy end of the funnel hung directly overhead. Everything was as still as death. There was a strong gassy odour and it seemed that I could not breathe. A screaming, hissing sound came directly from the end of the funnel, which was about 50 or 100 feet in diameter, and extending straight upward for a distance of at least one half mile, as best I could judge under the circumstances. The walls of this opening were of rotating clouds and the whole was made brilliantly visible by constant flashes of lightning which zigzagged from side to side. Had it not been for the lightning I could not have seen the opening, not any distance up into it anyway.

Around the lower rim of the vortex small tornadoes were constantly forming and breaking away. These looked like tails as they writhed their way around the end of the funnel. It was these that made the hissing noise. I noticed that the direction of rotation of the great whirl was anticlockwise, but the small twisters rotated both ways — some one way and some another.

The opening was entirely hollow except for something I could not exactly make out, but I supposed it was a detached wind cloud.

No meteorologist, not even the great Fujita, could have asked for a better vantage or a better description. The presence of inner vortices that Keller refers to as small tornadoes (which, in fact, they were) wasn't even guessed at until Fujita first speculated about their existence in 1965, but here is Keller describing them quite lucidly 37 years earlier. While the strong, gassy odor he reports has never been explained, the tubular cloud that snaked up the center of the tornado has since been witnessed.

No other storm exceeds the wind speed of a tornado. The debris field left in the wake of a major one often includes extraordinary testaments to a tornado's power and bizarre selectivity. Wheat straws are imbedded inches deep in tree trunks. An egg is found perforated with a neat hole that looked as if it had been drilled — inside is a bean. A diesel engine on the Union Pacific Railroad is lifted off its track, spun in midair and put back down again on a parallel set of tracks facing the opposite direction. A mattress with a child sleeping on it is sucked out the window of a Kansas farmhouse and dropped in a field nearby without even waking the child. A lit kerosene lamp is carried hundreds of yards and then set down upright with the flame still burning.

6

KATRINA

THE LIFE STORY OF A HURRICANE

"Y'all can't stop Mother Nature. If she wants to come 'n get 'cha, she'll come 'n get 'cha."

KATRINA SURVIVOR, NEW ORLEANS

I'm a born storm chaser. Growing up as I did in the middle of a summer storm zone at the southern end of the Great Lakes peninsula, I was witness every summer to hot Gulf air masses blowing north, up the Mississippi valley and spawning terrific thunderstorms over my hometown. I prayed for tornadoes. I stood outside in the wind before the rain, searching the blue-black clouds for a funnel, and, as an adult, I drove toward distant towering thunderheads, hoping to catch even the briefest glance of a twister. But my wishes were never fulfilled, at least not in the tornado department.

I'd arrived in Grand Bahama late Sunday afternoon on August 21, 2005. After unpacking, I took a sunset stroll along the beach. The air was hot, even in the early evening, and the ocean was calm enough for me to spot a dark patch in the water. Could it be a coral reef? After breakfast the next morning, I strapped on my mask, crouched in the seawater and pulled on my flippers.

The water was surprisingly warm and not only in the shallows. Just the mild exertion of a facedown breaststroke made me sweat, underwater, a decidedly peculiar sensation. (The purpose of perspiration — cooling the body — is defeated underwater, with sweat only pumping more salt into the ocean.) I had no idea that water this warm was unusual even for Grand Bahama in August and that it would play into meteorological events that were underway long before I landed in Freeport.

Three weeks earlier, around the time I was wondering what I should pack for my trip, the weather in North Africa was conspiring a chain of events that would turn my holiday upside down. And it all had to do with prevailing winds, the northern trade winds in this case, which in turn are entirely a result of the Coriolis effect.

THE CORIOLIS EFFECT

In the northern hemisphere, anything airborne and moving in a straight path in any direction — east, west, north or south — has a tendency to deflect to the right, while in the southern hemisphere it has the opposite inclination. This tendency, the Coriolis effect, is named after the French engineering professor, Gaspard-Gustave de Coriolis, who discovered it in 1835. It is a result of the Earth's spin, and the effect is so pervasive that even long-range artillery guns have to be specially calibrated to compensate; otherwise, they would miss their targets.

So why the difference between the northern and southern hemispheres? In both hemispheres, the sun rises in the east and sets in the west, but that's about it for similarities — because the hemispheres rotate in opposite directions. But how can this be? How can two halves of a single planet spin in opposite directions? It's one of the profound paradoxes of rotating spheres, but it can be solved with a simple thought experiment.

Imagine you are on a spacecraft approaching Earth from above the southern pole. You would descend on a planet spinning clockwise. But take the same spacecraft and approach the Earth from the North Pole, and you land on a planet spinning counterclockwise. Voila.

High-pressure systems in the northern hemisphere, whose outflow of wind is deflected to the right by the Coriolis effect, rotate clockwise. Northern hemisphere low-pressure systems cause an inflow of wind and so rotate counterclockwise. Just to confuse everything, meteorologists call high-pressure systems (which spin clockwise) anticyclones and low-pressure systems (which spin anticlockwise) cyclones. At least in Australia, their high- and low-pressure systems are less confusing: their anticyclones spin anticlockwise.

The Coriolis effect decreases with latitude. The closer you get to the equator, the less it deviates. Within five degrees above and below the equator, its effect is almost zero. But it definitely has an effect on subtropical weather beyond 15° north of the equator: it delivers hurricanes that hit the Caribbean region and North America.

This is what happens. Air over the equator heats, rises and is pushed north in the upper atmosphere. There it cools and sinks and travels south in the lower atmosphere, forming a kind of tubular vortex — a Hadley cell — that rings the Earth. The southward traveling winds of a Hadley cell are deflected to the right by the Coriolis effect, producing winds blowing from the northeast to the southwest. These are the northeast trade winds.

The northeast trade winds dominate a world-girdling belt of

ocean and land just north of the Tropic of Cancer, shunting all the weather in their path — highs, lows, tropical depressions — westward. They even blow red dust from the Sahara all the way to Florida. In the first week of August, when I was planning my trip to Grand Bahama, the northeast trade winds nudged an unusually violent storm system out of Chad through Nigeria, then on through southern Mali and Guinea. After drenching Guinea with torrential rain, the system blew out into the Atlantic Ocean just south of Cape Verde on August 8.

If North Africa is the nursery for fledgling hurricanes, then the Cape Verde islands are their finishing school. In the late summer months, these islands oversee the development of dozens of tropical waves (a term for a low-intensity storm) that then often transform into tropical depressions. Under the right conditions, tropical depressions turn into hurricanes. Just south of the Cape Verde islands, the African storm system that had left Guinea was coalescing into a tropical wave — a large area of low pressure studded with several active thunderstorms and sporting a complex, extensive cloud shield. Big enough to get some high-tech attention.

The National Hurricane Center has several satellites keeping an eye on the hurricane marshaling grounds. Aqua, launched in 2002, measures sea surface temperatures with a microwave scanning radiometer, and the TRMM (tropical rainfall measuring mission) satellite, a joint Japanese-American satellite launched in 1997, carries a lightning imaging sensor as well as a precipitation radar that gives scientists a three-dimensional view of rain within a storm. In satellite photographs, tropical disturbances look unorganized and patchy, like a random gathering of clouds. But some are a little swirlier than others, and in them you can vaguely see the outlines of a nascent hurricane, like an embryo in a sonogram. This was the case by August 11. The satellite data clearly showed the tropical wave had signs of convective organization, and two days later, on August 13, it was officially upgraded from a tropical wave to a tropical depression and given a number: 10. It was beginning to rotate, and wind speeds within the storm were reaching 25 to 38 miles per hour as it began its westerly drift along a strip of tropical Atlantic Ocean that stretches some 2,408 miles from the east African coast to the Caribbean.

HURRICANE ALLEY

During summer, the tropical Atlantic heats up like brine soup and the trade winds push developing hurricanes west. They grow larger and evolve from tropical waves into tropical depressions and then into tropical storms. Hurricane Alley is like the conveyer belt that runs through the oven of an industrial bakery, only the airy loaves that pass through Hurricane Alley's assembly-line bakery puff up on moist, thermal updrafts from a hot ocean. Water temperatures should be a minimum of 26.5°C down to a depth of at least 150 feet to properly spawn a storm. During a hot summer, Hurricane Alley can build a storm from its fledgling stage off the coast of Africa to a monster hurricane threatening the North American and Caribbean coasts.

Tropical depression 10 was 1,600 miles east of Barbados when it was given a number, but U.S. forecasters at the National Hurricane Center doubted it would get any bigger because mid-level wind shear had developed, likely to cap the ability of the storm to build a central core. Hurricanes are, in essence, warm-core systems that are stacked vertically. Hobbled by wind shear, tropical depression 10 meandered slowly west for a few more days until by August 14, a week before I arrived in Grand Bahama, it had degenerated and the National Hurricane Center had reclassified it as a tropical low pressure. It lost its number. By the 18th, it had dissipated almost completely.

This weak disturbance that was hardly worth tracking — the remnants of tropical depression 10 — would turn into Katrina, and her ability to surprise was about to become evident. Now a disorganized tropical low, it kept drifting westward, wafted by the northeast trades. It entered the ocean north of Puerto Rico a day after I landed in Freeport. There it joined forces with a local tropical low and formed an alarming alliance that caught the attention of the National Hurricane Center. Its QuikSCAT satellite, which measures near-surface wind speeds with a SeaWinds scatterometer (a microwave radar) capable of penetrating cloud cover, had picked up one of the key developmental stages of a hurricane: the lowering of the mid-level circulation to the water's surface. Wind-whipped ocean spray was feeding warm moisture directly into the hungry storm. On August 23, this new hybrid storm was given a new number, tropical depression 12, and it slid into the Bahamas late the same afternoon.

I had had two lovely sunny days in Grand Bahama before tropical depression 12 struck. On that Tuesday afternoon, the sky clouded over, and in the evening it began to rain. There was some wind but not a lot. During the afternoon, the National Hurricane Center had issued a tropical storm watch. There was a possibility that tropical depression 12 might intensify. The wind was increasing, and the local weather channel was clearly alarmed about the possibility of a tropical storm. Being used to the hysterics of my own local weather channels in Toronto, I was skeptical and figured I'd still be able to go snorkeling the next day.

When steady winds reach 39 miles per hour in a tropical depression, the National Hurricane Center gives it a name and upgrades its status to tropical storm. On Wednesday morning, August 24, tropical depression 12 was given the name Katrina. Well, I thought, at least a tropical storm isn't a hurricane, and although it was breezy in Freeport, it wasn't *that* windy. So I drove to the west end of the island to snorkel at my favorite offshore reef, Paradise Cove. There were few showers on the way, but the air was warm, and, more importantly, when I arrived the waves were not so large that the dive masters had shut down access to the outer reef. I parked, checked in, put on my flippers and mask and swam out with dark clouds looming ahead of me.

By the time I got to the outer reef, I was the only snorkeler and I could hear occasional thunder. The storm was definitely getting closer. Seeing lightning flicker under water, reflected from above, seemed totally worth the risk I was taking. It lent a dramatic, moody ambience to the electric colors of the fish and coral. I finished a circuit of the reef as the wind had begun to pick up. I could hear it whistling in my snorkel. As I came around the southern end of the reef, I noticed a motorboat heading toward me. The driver turned out to be one of the staff from Paradise Cove. "Do you need a ride?" he shouted over the wind. I lifted my mask while treading water and said, "No thanks, I'm fine." Though actually, it was a bit of a slog getting back to shore. The waves *were* getting a bit choppy.

On the drive back, I had to negotiate some fairly deep, large puddles — there are no storm sewers on Grand Bahama. Otherwise it didn't seem very stormy. Late that evening, the local television forecast showed radar images clearly revealing that tropical storm Katrina, which was just south of Grand Bahama, had broken up into three distinct fragments. As a result, the weather advisory was lifted. I went to bed just before midnight on a completely calm evening.

No one was sleeping at the National Hurricane Center though. High above me, careening through the upper thermosphere in the early morning hours of August 24, the NHC's TRMM satellite was sending some alarming data back to the NHC command center. The radar was detecting the formation of eyewall clouds south of Grand Bahama. This is where the heaviest rainbands (continuous, linked chains of thunderstorms) concentrate during the transformation from tropical storm to hurricane. At around three in the morning, Katrina opened its eye.

The formation of a hurricane's eye is still not completely

understood but the basic mechanics are well-known. A hurricane is a vast heat-sucking machine sustained by strong updrafts. As the rainbands begin to rotate around the hub of the hurricane, a strong ring of vertical convection forms just outside the midpoint. This massive circular updraft of warm air rises to the top of the hurricane, where it creates a layer of high pressure directly above the low-pressure heart of the hurricane. Most of this rising air flows outward over the hurricane in a clockwise spiral. This strengthens the vertical updraft, and consequently the power of the storm, in a mighty feedback loop. But in the small, restricted central core of the high-pressure lens above the center of the hurricane, the air is forced to flow inward. Without anywhere else to go, it begins to descend, creating a rainless, cloudless hole at the hurricane's center. The eye. Here, barometric pressure is at its lowest.

Generally speaking, the smaller the eye, the more powerful the storm. Hurricane Wilma, which followed Katrina in October 2005, had an eye only 2.3 miles across. It was an intensely destructive category 5 storm. Wilma set another record, one for low pressure, bottoming out at an ominous 26.05 inches. (Household barometers only go down to 27.5.) At the other end of the spectrum, Typhoon Carmen, which hit South Korea in 1960, had an immense 230-mile-wide eye and winds so low it barely qualified as a typhoon. I think you could've forgiven people in the path of Carmen for thinking that the storm was finished; it took hours for its eye to pass over them.

Typhoons are just hurricanes that form in the Pacific Ocean west of the international dateline. A hurricane that forms over the Indian Ocean is called a cyclone. Hurricane John, a Methuselah among hurricanes, lasted 31 days in August and September 1994. John crossed the international dateline twice, becoming Typhoon John and then doubling back to become Hurricane John again. If a hurricane skips from the Atlantic basin into the Pacific basin, like Earl did in August 2004, the name has to change. Earl became Frank.

I was woken up around four on Thursday morning by a repetitive banging sound. Someone must have come home late, but what were they doing? I got out of bed and raised the blind to look into the parking lot. Bedlam. Horizontal wind-whipped rain. The banging sound was coming from a metal rain gutter flapping against the ceramic roof tiles as the wind pried it off the building.

The word hurricane is derived from the Caribbean Arawak name for their storm god, Huricán, adopted by Spanish colonialists as *huracán*. Peak hurricane season for the West Indies, Bahamas and Florida is from mid-August to late October; during the thousands of years the Arawaks inhabited the Caribbean before the Europeans arrived, they were right in line for most of these storms.

Not all hurricanes are the same. None of them are minor, and some are formidable. As it turns out, I had picked a dilly of a year to visit Grand Bahama. The year 2005 broke all records with an unprecedented 15 hurricanes in one season. Four were category 5s.

CALIBRATING CATASTROPHES

The Saffir-Simpson scale measures a hurricane's strength. Herbert Saffir, an engineer from Coral Gables Florida, and Robert Simpson, then director of the National Hurricane Center, developed their hurricane-intensity measurement system in 1971. Saffir had devised the preliminary calculations earlier when he was commissioned by the United Nations to study the effects of hurricane wind on low-cost housing. Like Fujita before him, Saffir calculated the various thresholds of structural damage according to wind intensity. After some fine-tuning, he brought his 1–5 scale to Simpson, who added in the effects of storm surge and flooding.

The Saffir-Simpson scale was officially released by the NHC in 1973. Because it was so difficult to be accurate about such variable and sometimes immeasurable quanta as storm surge and flooding, especially after Katrina, these effects were removed from the scale in 2009. The new scale, based on wind speed only, became operational on May 15, 2010.

For a tropical storm to be promoted to hurricane status, it must have sustained winds of 74 to 95 miles per hour. At this intensity, category 1 on the Saffir-Simpson scale, the storm causes minimal damage. A category 5 hurricane has sustained winds of more than 155 miles per hour and causes catastrophic damage. Category 5 is generally considered the most extreme intensity possible, but on August 14, 1992, Hurricane Andrew exceeded the wind speeds of a category 5, hitting the southern Florida coast with 175 mile-per-hour winds. Andrew is one of the few examples of a superhurricane, unofficially designated as a category 6. It killed 65 people and flattened more than 63,000 homes.

Looking out over the rain-swept parking lot shining beneath the sodium security lights, I was, as they say, totally pumped, but also a little apprehensive. My building was solid, but how intense was the wind going to get? I went out front and stood on the balcony overlooking the beach. In the darkness, all I could see was the sideways glitter of rain in the security floodlights. Fortunately the wind direction was exactly parallel to the building and the beach. That eliminated the danger of storm surge for now, which was good because storm surge is the deadliest part of a hurricane, causing nine out of 10 hurricane-related fatalities.

Hurricane winds push water ashore, and if a hurricane's landfall happens to coincide with a high tide, the effect is the same as a tsunami. I have seen footage of a storm surge smashing into an already flooded parking lot and pushing cars and trucks ahead of it like so many pop bottles. A category 1 hurricane can create a storm surge of four to five feet above the normal tide line, while a category 5's surge can reach 18 feet.

One of the highest storm surges on record occurred in Bathurst Bay, Queensland, Australia, on March 5, 1899. A late-season tropical cyclone by the name of Mahina drove a 42-foot storm surge onto shore. If you were vacationing in Bathurst Bay that week, your ocean-view hotel room had better have been on the fifth floor or higher, because the bottom four stories would have been entirely flooded. To put it in perspective, the Boxing Day tsunami in Thailand swamped the bottom two floors of the seaside resorts there.

A power outage was pretty much inevitable, so while the water was still running, I filled containers from the kitchen tap. I completely forgot to fill the bathtub and sink, probably because I was rattled and tired. I went back to bed but hardly dozed off. The sound of the wind seemed to be increasing.

At first light, I checked the light switches: the power was off. Then I went to the front balcony (thank goodness I was on the second floor) to look at the storm. The wall separating my balcony from my neighbor's provided a natural wind break. The wind was blowing east to west — left to right for me — and as I watched, it flattened a brand-new picket fence that ran beside the oceanfront garden. Thirty-pound palm fronds that had been ripped from coconut trees were skidding along the beach like runaway boogie boards. The waves on the ocean beyond the beachfront garden were high but running parallel to the shore. No storm surge yet.

There was nothing for it — I had to go out and experience the storm directly. I was wearing shorts and a T-shirt and clutching the railing of the staircase leading down from my unit. I stepped out into the wind. It was like having your head out the window of a car on the highway, only in this case, my whole body was being blown. I descended the stairs with some difficulty because I had to lean into the wind. Walking to the beach, I employed the same leftward lean. There, on the shore I took in the spectacle. High waves running parallel to the shore were having their spume sheared away by the gale.

The wind caught at my face, and I could feel my skin rippling. The rain hurt. I thought at first it was just the velocity of rain, as if I was being hit by a water cannon, but then I realized it was filled with sand grains. It stung, viciously. So I did what anyone else would do, I leaned into the wind at an angle so steep that if the wind dropped, I would have fallen face first onto the wet sand. I've never experienced such constant, intense wind. There was something industrial about it, nothing like the variable gusts of a summer storm. It was more like standing in a wind tunnel.

Joseph Conrad wrote about hurricane wind in his book *Typhoon*, "This is the disintegrating power of a great wind: it isolates one from one's kind. An earthquake, a landslip, an avalanche, overtake a man incidentally, as it were — without passion. A furious gale attacks him like a personal enemy, tries to grasp his limbs, fastens upon his mind, seeks to rout his very spirit out of him."

Good enough. I headed back in and toweled off.

Katrina, as I experienced her that morning, was still a category 1 storm, but imagine what a category 5 would have felt like. I'm quite sure I wouldn't have been able to stand on the beach. Actually my metaphor of sticking my head out the window of a car could be extended but, in the case of a category 5 hurricane, instead of a highway think of a Formula One racetrack. Think of a house careening down an Indy 500 course at 155 miles per hour. The house's asphalt shingles wouldn't last through the first turn. Hurricanes can even peel off heavy, clay roof tiles. In fact, if you parked an Indy 500 car by the beach during a hurricane, it would sandblast the high-gloss enamel paint down to the metal. Hurricanes peel corrugated tin from roofs and then lift the whole roof off. They blow out windows, doors and completely flatten simple, wooden-frame buildings.

My position, I found out later, was 40 miles north of Katrina's eye as she headed for Florida. The phrase "headed for Florida" is a bit of a misnomer; she was already there. Most hurricane cloud shields are at least 300 miles in diameter, but it's only the eye that the National Hurricane Center tracks.

Katrina struck just south of Miami at half past six that evening, after which she veered south over the Everglades and by the next day, Friday, she was heading west into the Gulf of Mexico.

The power came back on in my building early Friday morning, August 26. The sun was starting to shine, and the wind had died down to an intermittent breeze. I drove to Freeport to get some groceries. There was some local flooding — I had to drive through puddles the size of ponds — but aside from tree branches and coconut palm fronds everywhere, Freeport looked intact. I appreciated, for the first time in a purely practical way, the stolid, neoclassical architecture of the government buildings: limestone columns and thick, limestone block walls. In fact, the building I was staying in, which had seemed overbuilt to me with its thick concrete walls and heavy roof tiles, now made perfect sense. What wasn't bolted, cemented or tied down would blow away.

Katrina was a nocturnal hurricane. She intensified in the darkness over warm seas, and the Gulf of Mexico was very warm that summer. She ramped up her power again around five on Saturday morning; before noon that day, she was a category 3 hurricane and looked to be headed directly toward New Orleans. Things were getting very serious. The state governor had issued a hurricane watch the day before, and now the mayor of New Orleans issued a voluntary evacuation recommendation.

Early that afternoon, a Lockheed Martin WC-130J aircraft stuffed with meteorological instruments took off from Keesler Air Force Base in Biloxi, Mississippi, piloted by the famous Hurricane Hunters, an elite Air Force Reserve unit of the 53rd Weather Reconnaissance Squadron. Following their usual hurricane flight plan, they made two penetrations into the eye of Katrina, one at 10,000 feet and another mind-boggling pass at a mere 500 feet, just above airborne palm fronds and roof tiles.

It was worse than anyone had thought. Katrina had grown into a monster. The Hurricane Hunters reported that its circulation covered the entire Gulf. Although Katrina was a category 3 hurricane when it hit New Orleans a day and half later, its storm surge was extraordinarily large, a catastrophic 25 to 28 feet in height. This was why the levees in New Orleans didn't have a chance.

On Sunday, August 28, while Katrina was approaching New Orleans, I flew back to Toronto. The only evidence of Katrina's passage through Miami, at least at the airport, were some puddles on the runways. The Monday morning news was already filled with scenes of chaos from New Orleans. The storm was a national disaster.

In a strange twist, after Katrina left New Orleans, she had one more punch left in its bag. She entirely destroyed the Keesler Air Force Base on her way north. It almost seemed vindictive. The base has since been rebuilt, and the Hurricane Hunters fly out of Biloxi once again.

Looking back, it seems to me that Katrina and I took turns stalking each other. First, she snuck up on me while I slept, then she abandoned me for Florida. I followed her a few days later, but she was already churning through the gulf. And on Wednesday the next week, she came all the way to Toronto.

In the interim, I followed Katrina's progress on the weather channel. She swept northeast on Tuesday, through Mississippi and then Tennessee where she was downgraded to a tropical depression. She still had a lot of momentum though, and the edge of her cloud shield slipped over Toronto that evening. The next day, the center of the storm absorbed a frontal boundary and became an extratropical storm, swirling through Ohio and across Lake Ontario into Toronto.

On the afternoon of Wednesday, August 31, I went into my backyard and stood in Katrina's warm deluge once more, six days after I had felt the sting of her gritty wind-whipped rain on the beach in Grand Bahama. Her fury had quieted to occasional gusts, but she still held a lot of rain. I let myself get soaked. Was I imagining it or was there a faint ocean scent in the air?

7

PALACE OF THE WINDS

"Who has seen the wind?
Neither you nor I:
But when the trees bow down their heads,
The wind is passing by."
CHRISTINA ROSSETTI

"Listen to no one's advice except that of the wind in the trees."
CLAUDE DEBUSSY

Wind is elusive, capricious, sensual and dangerous. From out of nowhere, a zephyr can brush your cheek as lightly as a feather or, with a sudden tug, a gust will turn your umbrella inside out. A scent can be carried on the wind to our nostrils, but wind itself is odorless, tasteless and colorless. Yet for all its transparency, it does have dimension. And sound. Wind sighs in the pine trees at night, and, as Rossetti witnessed, wind has shape — we see its muscular surges rippling fields of wheat and its intricacies in drifting smoke.

Each of us encloses an intimate wind — the narrow rushing

stream of air we inhale through our nose, which whistles down our throats into our lungs and then out again. The breath of life. That is why the ancients believed, as Anaximenes wrote in the sixth century BCE, that the world itself breathes and that the atmosphere was the *spiritus mundi* — the soul of the world. And the wind, which the ancient Greeks called *pneuma*, was heaven's breath.

The connection between wind and breath is everywhere in our idioms. Casual conversation is called *shooting the breeze*, but if you're an empty-headed gabber, you're a *windbag*. The answer may be *blowing in the wind* but only if you know *which way the wind blows*. And whatever you do, don't *throw caution to the wind* because then you'll be *sucking wind* or, worse still, find yourself *pissing into the wind*. Then when *the wind is taken out of your sails* and you try to drown your sorrows, you'll finish the day *three sheets to the wind*. Perhaps later you might *get wind of something*, maybe a potential *windfall*, but when you *run like the wind*, you might just end up trying to *catch your wind*.

But what is wind? In the fifteenth century, Leon Battista Alberti took it upon himself to summarize what was then known. Alberti epitomized the Renaissance man. Not only was he an author, a poet, a linguist, a philosopher, an artist, an athlete (he could leap over a standing man and ride wild horses) and a renowned architect, he was also somewhat of a natural historian. In his treatise on architecture, *De Re Aedificatoria*, in a sort of disclaimer near the beginning, he lists all the competing contemporary theories about the origin and nature of wind.

> *Whether these Conjectures be true, or whether the Wind be occasioned by a dry Fumosity of the Earth, or a hot Evaporation stirred by the pressure of the Cold; or that it be, as we may call it, the Breath of the Air; or nothing but the Air itself put into Agitation by the Motion of the World, or by the Course and Radiation*

of the Stars; or by the generating Spirit of all Things in its own Nature active, or something else not of a separate Existence, but consisting in the Air itself acting upon and inflamed by the Heat of the higher Air; or whatever other Opinion or Way of accounting for these Things be truer or more ancient, I shall pass over as not making to my purpose.

A century later, in 1563, the English meteorologist William Fulke declared that that wind is "an Exhalation hot and dry, drawn up into the Air by the power of the Sun, and by reason of the weight thereof being driven down, is laterally or sidelong carried about the Earth." I'm not sure exactly what that means, but I like it. The title of his publication is almost better: *A Goodly Gallery with a Most Pleasant Prospect, into the Garden of Natural Contemplation, to Behole the Natural Causes of All Kind of Meteors.*

More than 1,500 years earlier, Vitruvius, who had inspired Alberti, and whom we'll encounter again, wrote that wind is "a flowing wave of air, moving hither and thither indefinitely. It is produced when heat meets moisture, the rush of heat producing a mighty current of air." I like his elusive definition, the "hither and thither indefinitely." It captures well the whimsical nature of wind.

The most accurate explanation of wind is the simplist: it is air on its way elsewhere. It is the result of atmospheric pressure differences, always flowing from high to low pressure along the path of least resistance. Wind is air on a mission, on its way to balance inequalities.

THE ANEMOI:
GREEK GODS OF THE WINDS

According to Greek mythology, Aeolus was ruler of the winds, and he kept the four powerful gods of the cardinal winds (the Anemoi), as well as the Aellai and Aurai (the nymphs of the breezes), inside the hollow interior of his floating bronze island called Aeolia. In some accounts, Aeolus was a warden, incarcerating the winds on his island prison. Other accounts described the island as a stable for the winds. Whenever he released them, they thundered up through the peak of the island into the sky like the equestrian spume of some invisible volcano before they swooped down to scour the waves and land.

The Spartans, who loved horses, attached some significance to this association between wind, horses and mountains. They used to sacrifice a horse to the Anemoi on Profitis Ilias, the tallest mountain in the Taygetus mountain range in southern Greece. To claim that a horse ran as "fast as the wind" must have already been a common saying.

The north wind was one of the fastest. Of all the winds that Aeolus shepherded, it was Boreas, god of the north wind and winter, who was a favorite son of the Greeks. He was portrayed as a strong old man with flowing gray hair and a harsh temper. Boreas was also associated with autumn and was sometimes referred to as the "devouring one." He could, if he wished, take on the form of a horse. Pliny mentioned that mares that stood with their hindquarters to the north wind could bear foals without a stallion.

Because Boreas kidnapped an Athenian princess and fathered her children, the people of Athens considered him a relative by marriage and prayed to Boreas when the Persian army threatened Athens. He was a sly, lustful god. When Zeus abducted Europa, Boreas blew her tunic, lifting it up so that he could thrill at the sight of her lovely breasts. Yet his glimpse only served to inflame his jealousy that Zeus should carry off such a prize.

Zephyros was god of the west wind and bringer of spring and early summer. Like his brother Boreas, Zephyros was identified with old age and winter; he once had a sharp temper like Boreas, but his love for Flora, the goddess of the spring, mellowed his disposition. In friezes and mosaics, he is depicted gliding through the air, his mantle filled with flowers.

It is perhaps of the younger, more strident Zephyros that Shelley writes in his "Ode to the West Wind" (1819).

> *O wild West Wind, thou breath of Autumn's being,*
> *Thou, from whose unseen presence the leaves dead*
> *Are driven, like ghosts from an enchanter fleeing,*
> *Yellow, and black, and pale, and hectic red,*
> *Pestilence-stricken multitudes.*

In the Greek pantheon of wind dieties, only Notus, the south wind and bringer of rain and late summer storms, was young. His season was spring, and his dominion was sanguinous, for blood runs quickly in the spring. Any depictions of Notus portray him as a young man carrying an inverted water jar, spreading rain across the land. When he was airborne (like the rest of the wind gods, he often was), he flew wrapped in a cloud like a UFO in a Spielberg film. His realm was the tropics and North Africa, where hundreds of years later the greatest Roman city of all, Leptis Magna, rose from the desert in the dominion of Notus.

Eurus, god of the east wind, had no assigned season, although he was sometimes associated with summer and was thought to bring rain. He too was portrayed as an old man, though with a dark complexion and fierce features who often brought foul weather. In many depictions, he is crowned with a radiant sun, acknowledging that the new day rises in the east. His gloomy disposition belied his attributes, which were infancy and summer. And yet the east wind was reputed to be unlucky, particularly for seafarers. In *The Mirror of the Sea*, Joseph Conrad writes, "The East Wind, an interloper in the dominions of Westerly Weather, is an impassive-faced tyrant with a sharp poniard held behind his back for a treacherous stab."

The Greeks also deified the quarterly winds, of whom the god of the northeast wind, Kalkias, and the god of the northwest wind, Skiron, were yet again both intemperate old men. The god of the southwest wind, Lips, and the god of the southeast wind, Apeliotes, were young men.

As usual, the Romans put their own stamp on the Anemoi, although mostly it was only a shift of nomenclature. The god of the north wind, Boreas, became Aquilo, while the south wind, Notus, became Auster, from which Australia and the aurora australis are derived. For the Romans, Auster took on the additional responsibility for the sirocco, with its heavy cloud cover and humidity. Zephyros, god of the west wind, became Favonius, and the divisional winds were known collectively as the Venti, to which the Romans added Tempestates, the goddess of storm winds.

All of the Anemoi found themselves portrayed on one of the most singular monuments to meteorology ever constructed in the late classical period of Greece. It was a fusion of mythology, chronology and the budding science of forecasting.

THE ARCHITECTURE OF WIND

The Tower of the Winds, an eight-sided marble clock tower built by Andronicus of Cyrrhus in the second or first century BCE, still stands in the Roman Agora in Athens. It is 39 feet tall and 26 feet in diameter with a sundial on each face. When it was operating, it contained a water clock and was surmounted by a wind vane, a bronze figure of Triton whose trident pointed in the direction of the prevailing wind. (Wind vanes became very popular among well-heeled Romans. Within a century, the grandest villas in Rome boasted their own personal wind vanes, some of which connected to a pointer contained in a compass on the ceiling of the room beneath. The first meteorological instruments.)

The Tower of the Winds is aligned to the true points of the compass, with four sides of the octagon facing the cardinal directions and the other four sides bisecting them. Atop each of these facets is a sculpted depiction of the wind god that presides over that direction. They are winged, flying from left to right. We see Notus with his inverted water jug and Boreas blowing through a large shell. Lips holds the stern of a ship while a seminude Zephyros carries a host of flowers.

Vitruvius probably wrote *The Ten Books on Architecture* during the reign of Augustus, a hundred years or so after the construction of the Tower of the Winds. The exact date is disputed. Some say he wrote it during the reign of Nero. What *is* known is that he was an ambitious, talented architect, who courted the favor of Rome's patrician classes, particularly the Caesar, to whom *The Ten Books on Architecture* is dedicated.

In a chapter called "The Directions of the Streets; with Remarks on the Winds," he draws upon the writings of the Greek philosopher Hippocrates, who, five centuries earlier, wrote that towns or cities "that lie toward the risings of the sun are likely to be healthier than those facing the north and those exposed to the hot winds" and that their "inhabitants are of better complexion and more blooming than elsewhere . . . They are clear-voiced, and with better temper and intelligence than those who are exposed to the north."

In a similar vein, Vitruvius maintained that "cold winds are disagreeable, hot winds enervating, moist winds unhealthy." This is entirely in keeping with Italy's long and fraught relationship to draughts, winds and humidity, which has come down to us today virtually unaltered. I've taken train rides in Italy on unbearably hot days where not a single window in my coach could be opened for fear of someone getting sick from the draught. Vitruvius goes on to use the ill-aligned town of Mytilene, on the island of Lesbos in Greece, as an example of bad planning. It looks like a lovely place, but the positioning is all wrong: "In that community when the wind is south, the people fall ill; when it is northwest, it sets them coughing; with a north wind, they do indeed recover but cannot stand about in the alleys and streets, owing to the severe cold.

VILLA OF THE WINDS

The caves of Costozza in the Berici Hills of Vicenza have been celebrated by poets as sources of inspiration and divine insight since Roman times, but it was only during the Renaissance that Costozza's caves realized their climate-control potential. An opportunistic architect connected them to the adjacent Roman quarries and created an ingenious underground network of tunnels and caves. These formed a system of ventilation ducts that supplied three villas — Villa Eolia, Villa Trento-Carli and Villa Trento-Buoni Fanciulli — with the Renaissance version of central air.

The architect in question was Francesco Trento, who built the first of his three villas, Eolia, in 1560. It still stands in Costozza, an unremarkable, squarish edifice that is rescued by its more opulent interior — a large hall surmounted with a domed, frescoed ceiling. On the floor in the center of the room (now a restaurant) is an ironwork grate through which the wind rises from the subterranean networks. The grate forms the oculus, or window, on the ceiling of an eight-sided chamber beneath the hall (currently the restaurant's wine cellar), which contains eight empty niches in each of the walls. Just beneath these niches are inscribed the names of the local winds: Borea, Euro, Sirocho, Austro, Garbin, Zefiro, Maestro and Tramot.

In these names, we can decipher a mixture of Greek, Roman and local names. Borea might be a variation of bora, the cold, northeaster blowing out of the Alps; in Euro, we can recognize the Greek god of the east wind. Sirocho, of course, refers to the sirocco, the infamous southeasterly wind. Austro is probably a variant of Auster, Roman god of the south wind. Garbin is more mysterious, because the garbin is a Spanish southwesterly wind that blows from the Atlantic whilst the libeccio is an Italian southwesterly wind coming from the Mediterranean. Perhaps the Spanish term was an artifact from the period of Spanish rule. Zefiro, reminiscent of Zephyros, is probably the west wind of the western Adriatic, and Maestro is a strong summer northwesterly wind that blows in the same region. Tramot is probably a derivation of tramontana, the cold Alpine north wind.

Trento elaborated the natural air-conditioning theme with his next construction, the Villa Trento-Buoni Fanciulli, which not only had an underground wind channel but also a network of vertical conduits embedded within the walls that conducted cool, moist air to various rooms. These remarkable innovations were the architectural wonders of the time. "As I have experienced, nothing but a Terrestrial Paradise is sensed in these freshly ventilated rooms," wrote Francesco Barbarano, a priest and historian who chronicled the wonders of Trento's villas. They did more than refresh, however: they also embodied the Renaissance's architectural fascination with pneumatology, the science of enhancing the vitality of air and, along with it, the human spirit. Trento's Eolia and Fanciulli were *ville spiritali*.

The classical principles of proportion and elegance combined with the science of air (*pneuma*) and wind (*spirit*) saw their Renaissance apogee in the Villa Rotunda, completed by the architectual genius Andrea Palladio in 1566. Situated atop a hill, with four grand loggia and pediments facing the cardinal directions, the Villa Rotunda recalls Andronicus's Tower of the Winds. The Villa Rotunda had a cooling chamber à la Trento beneath the central rotunda that vented upward through a grate. The cool air then rose through the room, exiting via an oculus at the peak of the rotunda. This passive central air system provided constantly flowing cool air in the summer for the lucky owners of Palladio's architectural masterpiece.

FAMOUS WINDS:
FROM THE FÖHN TO THE SIROCCO

"It has now blown for these six days without intermission; and has indeed blown away all our gaiety and spirits; and if it continues much longer, I do not know what may be the consequence."

PATRICK BRYDONE,
SCOTTISH TRAVELER, 1776

One of the directional winds inscribed on the eight-sided apse beneath the floor of Villa Eolia is notorious. It is the Siroch, known today as the sirocco. If Trento had built his pneumatic edifice a little further north, in the Alps, then another local wind might have usurped the infamous sirocco, namely the föhn, a regional wind with a fearsome reputation and a complex genesis.

When large, humid air masses are pushed up the windward sides of tall ranges, such as the Alps or the Rockies, the mountains wring out every drop of moisture. This is because air expands as it rises in the lower pressure of altitude. In the thinner air, the molecules spread out and vibrate more slowly. But the molecules need energy to expand, and so the air cools in spite of the fact that no heat has been removed from it. It's a sleight of hand of physics called adiabatic cooling. When the air is moist, the cooling process is slowed down. When it's drier, adiabatic cooling speeds up.

The side effect of adiabatic cooling is rain and snow. Any moisture in an air mass that is pushed up the side of a mountain condenses, forming clouds that then release precipitation. A moist air mass will wreath a mountain summit in rainy gloom, called orographic rainfall. Under special conditions, after being thrust over the top of the mountains, the dry air rushes down the leeward side. It recompresses as it sinks and gains heat in a reverse adiabatic reaction that is amplified by the dryness and rapid descent of the wind, sometimes warming by 10°C for every 3,281 feet in altitude.

By the time this fast moving air hits the lower slopes of a mountain, it can be up to 40°C warmer than the ambient temperature of the valley.

In the Alps, this hot mountain wind is called the föhn, the "witch's wind." The föhn spawns wildfires and conflagrations in summer and melts snow in winter. It always seems to be accompanied by human complaints — migraines, insomnia and worse. Some hospitals in Switzerland and Bavaria postpone major surgery during a föhn because rates of postoperative deaths from heavy bleeding and thrombosis increase statistically at those times. Literally an ill wind.

In the Rocky Mountains, a similar kind of descending mountain wind, called a chinook, is famous for radical winter temperature swings. A chinook can be so dry and warm that it evaporates the snow in a process called sublimation (passing directly from a frozen to a gaseous state without melting first). In Lethbridge, Alberta, it is not unheard of for snow to evaporate at the rate of 12 inches per hour during a chinook. And the speed of the transition can be even more extraordinary. One day in January 1943, a chinook in the aptly named town of Rapid City, South Dakota, raised the temperature from -20°C to 7°C in less than five minutes. On January 15, 1972, a chinook took Loma, Montana, from -48°C to 9°C, a change of 57°C in less than 24 hours.

It seems meteorologists are inordinately fond of confusing monikers, particularly when it comes to related phenomena. Not only can you have adiabatic cooling and heating on mountains according to elevation, you also get anabatic and katabatic winds. An anabatic wind is warm wind that's blowing up the slope of a mountain and is caused by solar heating of the slope. A katabatic wind is one that blows air from a high elevation down a slope. The infamous Santa Ana winds of southern California are katabatic winds descending from the great arid plains northeast of the Sierra Nevada Mountains. The winds are so dry that, as they slide down the valleys of the Sierra Nevadas, they pick up heat as quickly as a föhn, making them blast furnace-hot by the time they hit Los Angeles and San Diego.

Santa Ana winds fan the flames of southern California's biggest wildfires. And like the föhn of the Alps, they also influence people's temperaments. In his short story, "Red Wind," Raymond Chandler wrote that on a night when the Santa Ana blows, "every booze party ends in a fight. Meek little wives feel the edge of the carving knife and study their husbands' necks. Anything can happen. You can even get a full glass of beer at a cocktail lounge."

Other mountain winds are not so warm. The cold, dry mistral that howls down the Rhone River valley, reaching speeds of 93 miles per hour, ruins holidays in the French Riviera and the Gulf of Lyon. It has been doing so for thousands of years. Around 10 CE, the Greek geographer Strabo called it "an impetuous and terrible wind which displaces rocks, hurls men from their chariots, breaks their limbs and strips them of their clothes and weapons." In contemporary times, it has been known to knock down brick chimneys, blow roofing tiles off homes and even topple railway freight cars.

Not all ill winds are caused by mountains. There is another wind, a native of southern Europe and the Middle East, which is conjured mostly in the spring by low-pressure systems moving eastward across the Mediterranean. These lows snare hot, dusty, air funneled up from the Sahara desert and the Arabian Peninsula, supercharge it with humidity and then blow it inland.

In Morocco, it is called leveche; in Tunisia, it is known as the ghibli; in Egypt, as the khamsin; in Iraq, it is called the shamal; in Israel, sharav; and in southern Europe, it is referred to by its Arabic name, sirocco. No one gives it good reviews. The Georgian-era travel writer Patrick Brydone devoted more than a page of his popular 1776 book *A Tour through Sicily and Malta, in a Series of Letters to William Beckford Esq., of Somerly in Suffolk* to the sirocco, talking about the lassitude, the headaches and the ennui it evokes in tourists. The local population is not immune either:

> *A Neopolitan lover avoids his mistress with the utmost care in the time of the sirocc, and the indolence it inspires, is almost sufficient to extinguish every passion. All works of genius are laid aside, during its continuance; and when anything very flat or insipid is produced, the strongest phrase of disapprobation they can bestow, is that it was written in the time of the sirocc.*

A French acquaintance of Brydone's complains, "'Ah, *mon ami*,' said he, 'I am near to death, I, who never knew the meaning of the word *ennui*. *Mais cet exécrable vent*, if it lasts even two days more, I will hang myself.'"

Two years before these lamentations were published, in County Meath, Ireland, Mary Beaufort, wife of Daniel Augustus Beaufort, gave birth to a boy who was destined to become master of the wind.

THE BEAUFORT SCALE

At first glance, it would seem that wind speed over water doesn't get the same respect as wind speed over land. Why is it measured in knots instead of miles or kilometers per hour? It's almost as if lakes and oceans render the exact velocity of wind a little more nebulous or approximate. Is the normal scale of wind speed no longer applicable near large amounts of water? What is it about water that makes such measurements subject to this strange conversion, like a foreign currency? Well, it all hinges on reference points. In order to measure wind speed accurately you need a fixed, immovable point, and that is hard to find in the open sea.

On water, everything is relative — your ship is moving in one direction, the waves are moving in another and the currents beneath your bow are moving in yet another. So, maritimers use the Beaufort scale, which, as it turns out, is an ingenious way of accurately measuring a vessel's speed.

Daniel Beaufort was a Protestant clergyman and a member of the Royal Irish Academy. He was a first-rate classicist, a scholar, a society man and a crack cartographer, publishing in 1792 the most accurate map of Ireland. Unfortunately, he was also improvident. His young family had to keep moving to avoid the bailiff. They relocated six times in Ireland and England during the first 16 years of his son Francis's life.

As a result, Francis's education was compromised, with one notable and critical exception. In 1788, at the age of 14, he attended classes led by Dr. Henry Usher, professor of astronomy at Trinity College Dublin. The classes were held inside the recently constructed Dunsink Observatory. Here he learned the nature of the stars, the constellations and planets. Through the great Dunsink telescope, he saw the crescent of Venus, surveyed lunar mountains and viewed the moons of Jupiter. He learned how to use a sextant to calculate his exact position on the global grid of longitude and latitude.

Usher was a spectacular influence on the young Beaufort, honing his powers of scientific observation. In a celestial journal that Francis kept at this time is a description of a lunar halo, a sure sign of impending rain.

> *On the 12th December, 1788 at a little after 11 o'clock I saw a circle around the moon at a distance of about 8' or 9' the breadth of it was a semi (diameter) of the moon it consisted of three shades, the internal one that next the moon was a lightish purple, next that a light red, and next a greenish yellow.*

A year later, Francis's father secured a berth for him aboard the *Vansittart*, a three-masted frigate bound for Batavia in the Dutch East Indies (present-day Jakarta). It was the inauguration of his naval career. Three weeks into the voyage, Francis was named the official midday latitude observer. Indeed, the 15-year-old was so proficient with a sextant that he revised the official cartographic position of the city of Batavia by three miles. He was correct.

Disastrously, a few days after leaving Batavia for the return voyage, the *Vansittart* struck a shoal that had been improperly marked on the nautical charts and sank. Beaufort survived the wreck and made his way back to England to enlist in the Royal Navy just in time for war with France. He rose rapidly through the ranks of the British navy; by age 22, he was appointed first lieutenant of the Mediterranean-based warship HMS *Phaeton*. There, in the fall of 1796, the *Phaeton* attacked and captured a Spanish brig. Beaufort was one of the first to jump aboard the enemy vessel, where he was shot at point-blank range with a musket, riddled with shrapnel from an exploding grenade and set upon by a saber-wielding Spaniard who delivered two heavy blows to his head.

Again he survived, but he was laid up for weeks with a musket ball in his left lung and more than a dozen other wounds. A few years later, in 1805, he was given his first command. His vessel, the HMS *Woolwich*, was assigned to conduct a hydrographic survey of the coast of South America. This was followed by many more hydrographic surveys, and it was during these expeditions that he developed his first versions of his wind force scale, as well as rough notes for his book on forecasting at sea eventually entitled *Weather Notation*.

In 1812, his Royal Navy surveys found him in the eastern Mediterranean commanding the HMS *Frederickstein* on a mission that also included patrols against local pirates. In June, a landing party from his ship was fired upon by pashas, and Beaufort went ashore to rescue his crew. As they were returning to the ship, a sniper shot Beaufort and the ball shattered his hip. He convalesced onboard for months and eventually returned to England, but he never saw active sea duty again, although he remained in the navy. In 1829, he was appointed hydrographer to the Admiralty; and in 1846, he was promoted to rear admiral. By 1838, the British navy adopted his wind force scale, which is still used by all seagoing vessels.

The standard unit for Beaufort's scale is knots, derived from an earlier system of calculating speed at sea by throwing a "chip log" overboard with a line attached to it. The line was knotted at uniform, measured distances and was played out for a fixed amount of time measured by a small hourglass. The number of knots that played out during that interval calculated the speed of the boat. Each knot is equal in speed to one nautical mile per hour, or 1.15 miles per hour.

Beaufort's knots measured wind speed in a series from 0 (dead calm) to 12 (hurricane force winds). A Beaufort number of 2 is a light breeze or a vessel underway at one to two knots, while a moderate breeze with a Beaufort number of 4 is equivalent to five to six knots. In the original Beaufort scale, anything over 4 was not measured by knots but by the effect of the wind on sails. A "moderate gale" (Beaufort 6) required double-reefed topsails and jibs, whereas a "strong gale" (Beaufort 9) required close-reefed topsails and courses.

Today's Beaufort scale has undergone a few modifications. In the early twentieth century, Antarctic explorers encountered such extraordinarily fierce winds in the blizzards howling around the South Pole that they added six more calibrations, bringing the scale up to 18. Those must have been some desperate measurements, because 12 on the original scale was at least 75 miles per hour, the strength of a category 1 hurricane. In Antarctica, the highest wind speed recorded was 200 miles per hour.

THE JET STREAM

The Second World War was the first conflict in which meteorologists played a strategic role. So much so that, in England, in a measure that inconvenienced many citizens, public weather broadcasts were halted for the duration of the conflict. At the same time that meteorology was becoming a hard science, the war was becoming a crucible for forecasting skills, and some of the best and brightest young meteorologists were enlisted by the U.S. military to provide accurate forecasts for battles in the Pacific arena.

One of them was Reid Bryson, a geologist and meteorologist stationed in Guam. Unlike many of his peers, Bryson was familiar with the work of Heinrich Seilkopf, the German meteorologist who, in 1939, claimed that he had discovered high-altitude winds, which he called "jet currents." Bryson was quick to adopt and use Seilkopf's theories. Bryson predicted that bombers flying toward Japan might encounter high-altitude westerly headwinds that would slow their progress. Anecdotal reports from pilots crossing the Atlantic had already mentioned high-altitude westerly tailwinds that boosted their ground speed by up to 100 miles per hour.

Bryson was right. A secret U.S. Air Force mission deployed to strike Honshu, an industrial target near Tokyo, flew into a jet stream in late November 1944. One hundred and eleven B-29 bombers with specially trained crews on their way to make the first high-altitude bombing run of the Second World War ran into disastrous winds. Everything went smoothly until, approaching their target, they turned and began their west-to-east run above Honshu at 33,000 feet. As they began to release their bombs, the whole fleet lurched forward as if pushed by a giant hand. They had flown into the path of a 402-mile-per-hour gale and most of their bombs missed the target and fell harmlessly into the sea.

Later, other high-altitude bombers on their way to targets in Japan ran into similar mysterious headwinds so strong that one of the pilots recounted how "islands that they should have passed long ago remained stationary below them as though the whole scene had frozen." Some missions were forced to turn back before they ran out of fuel.

These aviation hazards had to be understood and mapped out.

Over the next few years, meteorologists discovered that there were four tubes (wiggly planetary hoops, actually) of high-speed wind snaking along the top of the global tropopause like sidewinders in the sand — two in the northern hemisphere and two in the southern. Similar to rivers, their velocity was slower at the edges and highest in the core, where winds reached 310 miles per hour.

If wind is air on its way elsewhere, then the jet stream is a high-level express lane for the westerlies. Like serpentine free-floating wind tunnels, they separate the three major convection cells of the northern and southern hemispheres: the polar cell, the Ferrel cell and the Hadley cell.

An atmospheric cell is a giant toroidal vortex, like a toeless sock being perpetually turned inside out or, better still, like a smoke ring. If you could somehow get a smoke ring to hover horizontally and then very carefully inserted a small globe up into the rolling ring, you would have a small-scale version of the Hadley cell. (Sure, a cross section of the smoke ring would be circular, while a cross section of any of the atmospheric cells would be more elliptical, but the analogy is pretty close.)

The Hadley cell is formed by air rising straight up from the equator, moving north in the upper atmosphere and sinking at about 30°N latitude, the same approximate latitude as Morocco, Georgia, Cairo and Shanghai. (This sinking air is dry; it's the reason that most of the great deserts — Sahara, Sonora, Gobi — are also smack-dab on 30°N latitude.) Once it hits the Earth's surface, the air moves south toward the equator, completing the Hadley circle. The northeast trade winds are a result of that southward-moving air being deflected by the Coriolis effect.

The Ferrel cell is a circular conveyer of air exactly like the Hadley cell except it rolls in the opposite direction. The Ferrel rises from about 60°N — think Stockholm, Whitehorse and the Shetland Islands — moves south in the upper atmosphere and sinks at 30°N where the Hadley cell is also descending. The last of the three cells, the polar cell, rolls in the opposite direction to the Ferrel (the same direction as the Hadley), rising up from 60°N to move north and sinking at the North Pole. From there, it spreads out and moves south in the lower atmosphere. The Hadley, Ferrel and polar cells are mirrored in the southern hemisphere with the same names.

In a sense, these bands of rolling air are similar to a milkshake in a blender: the fluid rises up the side of the blender, then moves across the top to the center where it sinks down into the blades in a single vortex. This metaphor works perfectly for the polar cell but not quite as well for the other cells, which are bands. Each of their vortices becomes an elongated groove running the length of the Hadley/Ferrel milkshake — in this case, right around the world. Wherever that groove occurs, whether between the polar cell and the Ferrel cell or between the Ferrel cell and the Hadley cell, a jet stream sits directly on that border. That's why there are two of them in each hemisphere.

Now it gets a little more complicated, so we need to introduce another scientist: Carl-Gustaf Rossby. Born in Sweden in 1898, Rossby moved in 1925 to the U.S. to teach in the aeronautics department of MIT. Fifteen years later, at the beginning of the Second World War, he was appointed chair of the meteorology department at the University of Chicago (where Fujita later became a faculty member). Like the great American meteorologist Cleveland Abbe before him, Rossby was a fluid dynamics guy; he applied the same mathematical principles that described fluid dynamics to the atmosphere. Just before the war, in the same year that Heinrich Seilkopf discovered the jet stream, Rossby had made a momentous observation, one that forever changed forecasting and meteorology. He had discovered his eponymous waves.

The figure/ground relationship of jet stream and Rossby waves are central to modern forecasting. To explain what Rossby waves are, we have to return to the Coriolis effect: the Earth's rotation causes storms to move clockwise in the southern hemisphere and counterclockwise in the northern hemisphere, an effect that gets stronger as you approach the poles. The closer to the pole something is, the more aligned it is with the axis of rotation (the invisible rod through the center of the Earth that connects the North and South Poles). The stronger the Coriolis effect, the stronger the destabilizing shear boundary between the polar cell and the Ferrel cell becomes, introducing points and bays like a shoreline. These points and bays are the Rossby waves. They move slowly, sometimes in a westerly direction and sometimes in an easterly direction. The jet stream follows their contours, which is why they are so important to forecasting weather.

Many a meteorological student has wrestled with the complex math of Rossby waves, not to mention the labyrinthine mathematics of fluid dynamics. But forecasting the weather doesn't need to be that complicated all the time. It can be as simple as standing out in your backyard or the street on a windy day.

FORECASTING BY WIND

The direction of the wind tells us a lot about what will happen to the weather. It has to do with an insight of a Dutch physicist in 1857. Christophorus Henricus Diedericus Buys Ballot, known to posterity as Buys Ballot, taught at the University of Utrecht in Holland. In 1845, he famously confirmed the Doppler effect (the change in pitch created when a moving source of sound, say an ambulance siren, approaches or moves away from the listener) with a group of musicians playing in an open train car. His interests turned to meteorology shortly thereafter. The first weather maps showing barometric pressure gradients, or isobars (regions of equal pressure), were just beginning to be published, and Buys Ballot was fascinated. What caught his attention was not the wiggly bull's-eyes of high- and low-pressure areas, but the wind-direction arrows between the isobar lines. He noticed a pattern.

The wind always blew *along* these isobars, not across them, and if you factored in the rotation of high- and low-pressure systems, Buys Ballot realized you could predict the weather by standing with your back to the wind. So in the northern hemisphere, the area of low pressure would always be to your left and the high pressure to your right. (The reverse would be true in the southern hemisphere.) Given the fact that most weather in the mid-latitudes moves west to east, it becomes a simple matter to make a rudimentary, fairly accurate forecast.

For instance, if the wind is from the north, you're between a low-pressure area rotating counterclockwise to your left and a high-pressure region rotating clockwise to your right. As the high-pressure area moves over you, the weather will most likely change from rainy to clear and cooler. Buys Ballot intended his shorthand forecasting law to be a means to save lives at sea, which it was. Sailors were able to apply his general rule and avoid storms and hurricanes well in advance. For long-range forecasting, however, sailors had to wait another hundred years.

8
WHICH WAY THE WIND BLOWS
THE STORY OF WEATHER FORECASTING

"The Book of Nature is written in the language of mathematics."
GALILEO GALILEI

Our inability to see into the future has frustrated humans for thousands of years. Along with our evolutionary leap into self-consciousness came the realization that we were enslaved by time's arrow, that we're blindly groping toward the future unaware of the imminent catastrophes that might ambush us there. This is indeed a limit to our assumed omniscience. You might call it a state of temporal claustrophobia, and once it was an anxiety that could only be remedied by prophets and soothsayers. Every ancient civilization honored citizens who claimed they could predict the shape of things to come and it was on these few seers we laid our hopes.

The early Romans turned to augers for advice. The appearance and flight of birds were central to these prognostications. An eagle that landed on Augustus's imperial tent while he was on campaign foretold his victory at Actium. Far north of Rome, in Scandinavia, legend says that Odin sent out his two ravens, Huginn (Old Norse for "thought") and Muninn (Old Norse for "mind"), into the world every day to bring back knowledge. Odin was obsessed with soothsaying, and he ultimately sacrificed an eye to the Norns for the ability to see into the future. Actually, Odin got a deal. In Greek mythology, poor Tireseas lost sight in both eyes for the gift of prophecy, and in his case the trade-off was not voluntary. Athena blinded him but then took pity and granted him the ability to understand bird songs and their divinations. Even today, palm readers do a brisk trade and futurist speakers are hot tickets at investment seminars. But none of them hold a candle to meteorologists for accuracy, at least in the short term.

The seventeenth century was a golden age of prognostication. Once Galileo and Kepler had figured out the orbits of the planets, it was child's play to determine when alignments and conjunctions would occur. That's why astronomers can take the cake for long-term predictions, able as they are to predict eclipses hundreds of years into the future. But their divinations are based on the clockwork orbits of planets and stars, which is, ultimately, not that much different from glancing at a wristwatch and noting that six o'clock will follow five o'clock.

Predicting the turbulent and erratic vagaries of the atmosphere is of another order of magnitude altogether. A modern weather forecast is like pinning a will-o'-the-wisp to a wall, more akin to augury than astronomy. Yet even here, in the world of turbulence and butterfly effects, mathematics underpins the works. The ephemeral clouds, the fickle winds and variable ocean currents are all funneled through the analytical machinery of hydrodynamic and thermodynamic equations. But I'm getting ahead of myself.

The mathematics that describes weather was a long time coming. The fluctuating cycles of weather and climate are deeply wired into all living creatures, an instinctual knowledge that gives them the ability to anticipate changes in the weather with a sort of climatic prescience — from birds flocking into a leafy tree just before a storm to green darner dragonflies migrating south in the fall. We humans too have a visceral, deeply wired relationship with the weather. Something primal that underlies our quotidian observations. Migraine sufferers know all too well when the barometer plunges. But ultimately, it is is our ability to observe, remember and see patterns that allows us to build our collective body of knowledge.

No one knows just when the first forecasts were made by humans, but it was likely very early on in our development. Chimpanzees have been observed gathering to watch particularly colorful sunsets, so it's safe to assume we had an eye for certain regularities in the weather long before we ventured out of Africa, regularities that hinted what the skies held in store. When we transformed from hunter-gatherers to farmers, a foreknowledge of weather made the difference between survival and ruin, between wealth and poverty. A crop of mature wheat could be wiped out by too much rain, and knowing oncoming weather even a day or two ahead of time could save the lives of citizens in the coming winter. The color of the rising and setting sun was probably one of the first meteorological forecasts. We know it today as "red at night, sailors' delight; red in the morning, sailors' warning." Similar, written observations have been found that date back to 650 BCE, when the Babylonians studied cloud patterns.

But it was the Greeks, a few centuries later, who initiated the next stage in forecasting. It might be argued that they invented the first daily weather forecasts, which were posted on columns in the agora of many of their cities. These were known as parapegmata, or peg almanacs, but they were more like nowcasts, often consisting of accounts of local conditions sometimes accompanied by astronomical details, "Rising of Arcturus; south wind, rain and thunder." One of my favorites reads simply, "The weather will likely change."

A parapegmata was still in use in Athens when Aristotle published his *Meteorologica*, a compendium of speculations about the four elements in 340 BCE: "Fire, air, water, earth, we assert, originate from one another, and each of them exists potentially in each, as all things do that can be resolved into a common and ultimate substrate." He didn't have the details we have today, but he more-or-less sketched out the carbon cycle — where air (carbon dioxide) is locked in earth and ultimately released by fire (volcanoes) into the atmosphere. He also theorizes about clouds, wind and the formation of dew, among other phenomena. Some of it sounds eccentric, such as "When there is a great quantity of exhalation and it is rare and is squeezed out in the cloud itself we get a thunderbolt," but given that most of his contemporaries were convinced that Zeus was the one hurling lightning bolts, he was a radical logician.

Around the same time Aristotle published *Meterologica*, his pupil and eventual colleague Theophtrastus of Eresus assembled a compendium of folk weather sayings in *De Signis* ("Book of Signs"). These arcana have a rural flavor and show how rooted Greek civilization was in the natural history of the Aegean area: "It is a sign of rain if ants in a hollow place carry their eggs up from the ant-hill to the high ground, a sign of fair weather if they carry them down"; "It is a sign of wind or rain when a heron utters his note at early morning: if, as he flies toward the sea, he utters his cry, it is a sign of rain rather than wind, and in general, if he makes a loud cry, it portends wind"; and "If a lamp burns quietly during a storm, it indicates fair weather."

These adages were the *Farmer's Almanac* of the times, and some of them were reliably predictive. But there were other issues that Aristotle grappled with. A theoretical abstraction was making the rounds: under certain circumstances, in the absence of any material substance, there might exist something called a vacuum. Well, not *something* at all, really *nothing*. Aristotle dismissed the notion of nothingness and issued an assertion that remained unchallenged for almost two millennia — that a vacuum was a "logical contradiction."

Like so much else, the received wisdom of Greek folklore passed on to the Romans. Cicero, the great orator who could recite his speeches forward and backward, wrote, "Storms are often stirred up by some particular constellation." Pliny the Elder (who perished in Pompeii after commandeering an imperial Roman navy ship to take him to the eruption and confidently reminding a fellow passenger that fortune favors the brave) wrote copiously about weather in his *Natural History*: "When clouds sweep over the sky in fine weather, wind is to be expected in whichever quarter the clouds come from" and "Jellyfish on the surface of the sea portend several days' storm."

In 904 CE, the great Persian Muslim scientist Ibn Wahshiyya translated from Nabataean Arabic a book on agriculture, which claimed rain could be predicted using lunar phases and atmospheric changes. Five centuries later, Leonardo da Vinci assembled the first hygrometer to measure humidity. But the most important discovery for modern forecasting wasn't made until 1650 when Evangelista Torricelli invented the barometer.

THE INVENTION OF THE BAROMETER

In 1608, Galileo Galilei heard about a spectacle maker in Holland who had fashioned an arrangement of lenses with the remarkable power of making distant things appear closer. He reproduced the arrangement on his own in Pisa and then, seeing room for improvement, built the world's first astronomical telescope, turned it on the moon and became the first human being to see lunar mountains. Looking at Jupiter, he discovered that it was orbited by several moons. Everywhere he turned his telescope in the night sky was bristling with stars and nebulae, and everything in this newly immense universe confirmed Copernicus's heretical assertion that the Earth orbited the sun, not the other way around. There was a big party going on out there, and the Vatican was not invited. All this was plain for anyone with a telescope.

But Earth's atmosphere had yet to be fully understood. Air was thought to be weightless, and anything less than air was deemed to be impossible. Aristotle's proclamation that nature abhors a vacuum was still regarded as a fundamental truth when, in 1630, a young scientist named Giovanni Battista Baliani wrote to his mentor, Galileo Galilei, concerning a problem he'd discovered with a siphon he had constructed to conduct water. He was using a suction pump to force water uphill, trying to ascertain just how high it could rise.

Baliani found that if the elevation of the transfer siphon reached a critical height, somewhere around 34 feet, the water stopped flowing. Something impeded it. Hence his letter. Galileo responded (years later in a 1638 publication) with two extraordinary speculations: not only did he acknowledge that a vacuum could exist (in the empty space at the top of Baliani's siphon), he also asserted the vacuum wasn't quite strong enough to raise the weight of the water above the 34-foot level.

Two years later, this speculation inspired two young scientists, Gasparo Berti and Raffaello Magiotti, to fill a 42-foot-long lead pipe with water and plug both ends. They then stood the tube in a water-filled basin and opened the bottom plug. Counterintuitively the water in the cylinder didn't entirely flow out. In fact, hardly any water spilled into the basin; the level inside the pipe had dropped only slightly. Because the upper end was sealed, there had to be a space at the top of the cylinder. What could be in that space? Had Berti and Magiotti succeeded in producing the very thing that nature abhored, something unnatural and yet so powerful it held up the column of water?

Evangelista Torricelli, a 32-year-old mathematician from Rome, applauded the experiment — the creation of the first generally recognized artificial vacuum — but he was more intrigued by the height of the water remaining in the tube. Berti and Magiotti estimated the level to be about 34 feet, the same as the height in Baliani's tube. Why the consistency? In 1631, nine years earlier, René Descartes had theorized that air might have weight and further speculated that it might be possible to construct a device to measure it. Torricelli realized that Berti and Magiotti's experimental apparatus *was* that device — it was showing that the weight of the atmosphere *pushed* the water up the tube and prevented it from running out. Torricelli had discovered the barometer. At least in theory. All he had to do was build his own.

Back in his laboratory, Torricelli set about constructing a prototype. But the Inquisition was raging just outside his laboratory windows, and he had nosy, gossipy neighbors. A two-story barometer erected in his courtyard would be viewed as the devil's work. Scientists were considered heretics to be locked in dungeons or burned at the stake like Giordano Bruno. Even the great Galileo, court mathematician to Ferdinando II, Grand Duke of Tuscany, had been forced to publically recant his scientific discoveries and go into exile. (Though Galileo's banishment to his villa in Arcetri hardly seems punitive.)

Unfortunately, the ceilings in Torricelli's laboratory weren't high enough to accomodate a 34-foot barometer. He needed to find a replacement for the water, something denser that would permit him to scale down the barometer's size. Mercury fit the bill. It was 14 times denser than water, allowing the 34-foot lead pipe to be replaced with a 32-inch glass tube. Once the mercury found its level, Torricelli could *see* the airless gap at the sealed top of the tube. Best of all, the whole apparatus fit easily on a small desk.

In science, it seems the important stuff is always invisible at first: radiation, radio waves, gravity and the weight of air. Now the unseen and impalpable realm of atmospheric pressure was made visible. In late spring 1644, Torricelli wrote to his friend Michelangelo Ricci and declared, "We live submerged at the bottom of an ocean of elementary air, which is known by incontestable experiments to have weight." Not only was the unseen realm of atmospheric pressure now visible, it was now measurable. As well, Torricelli calibrated the column of mercury and documented the daily variations in its height. He wrote that his instrument "will exhibit changes in the atmosphere, which is sometimes heavier and at other times lighter and thinner."

With his barometer, Torricelli had an observable vacuum with which he could perform other experiments. In one of them, he tried to determine if sound traveled in a vacuum, but the results were inconclusive. He also, curiously, placed insects in the vacuum to see if they survived. Insects don't look like they breathe, but as Torricelli undoubtedly discovered, they do.

Torricelli's mercury barometer underwent many refinements over the next few centuries. Blaise Pascal developed a portable one in 1646 and two years later used it on a mountain to confirm the decrease in air density with increased altitude. Manufacturers got into the game. Everyone was curious about these little forecasting devices, and by 1670, the first year of the coldest period of the Little Ice Age, barometers became the must-have conversation piece for the wealthy. As the price came down, the craze spread to the middle class. Over the next 200 years, the market for barometers went global, with more than 3,000 barometer manufacturers registered in Europe. The era of personal forecasting had arrived. When your barometer plunged, you knew bad weather was on the way. In its ornate, hand-carved wooden case, the barometer was a household wonder.

Scientists continued to refine the barometer as well as its

sibling, the hygrometer, invented by Leonardo da Vinci and perfected in 1783 by Swiss physicist Horace Bénédict de Saussure. A hygrometer measures the water vapor in the atmosphere, and de Saussure's version used a single human hair to do so. (Stretched taut in a brass frame, the hair was connected to a needle indicator on a scale — changes in its length caused by humidity were thereby measured.) De Saussure bemoaned the fact that all the scientific progress in weather forecasting was still no better than folk wisdom — "When there is enough clear blue sky to patch a Dutchman's breeches, expect fair weather," as one of the folk sayings went. He wrote, "It is humiliating to those who have been much occupied in cultivating the Science of Meterology, to see an agriculturist or a waterman, who has neither instruments nor theory, foretell the future changes of the weather many days before they happen, with a precision, which the Philosopher, aided by all the resources of Science, would be unable to attain."

In 1843, a French physicist, Lucien Vidie, built an aneroid barometer. This clock-like instrument didn't use mercury. Rather it relied on a small metal drum containing a partial vacuum that expanded and contracted as the air pressure went up and down. The drum was attached to a spring and, through a series of levers, to a needle on the face of the barometer. Aneroid barometers were smaller and more robust than mercury barometers. A year later, Vidie also perfected the barograph, which replaced the needle on the barometer with a pen that inscribed a line on a roll of paper that covered a rotating, clockwork drum. The resulting graph, rising and falling over a period of several days, became a permanent record of changes in air pressure.

WEATHER FORECASTS

With the hygrometer, barometer and thermometer now all in use, the stage was set for the first public weather forecasts, and their architect was Robert FitzRoy (1805–1865), the son of the third Duke of Grafton and great-grandson of Charles II. He grew up in a Palladian mansion in Northamptonshire and as a child dreamed of sailing with the British navy and commanding one of the great three-masted schooners that were then the mainstay of the fleet. To this end, at the age of 12, he enrolled with the Royal Naval College and two years later joined the Royal Navy as a deck hand, sailing on the frigate HMS *Owen Glendower* on its six-month journey to South America.

Upon his return in January 1822, Robert regaled his family with stories of exotic lands. Otherwise, 1822 was not to be a good year for the FitzRoys. Robert's uncle, Viscount Castlereagh, was the British foreign secretary, but he had had a few bad years. Lampooned in a poem by Percy Bysshe Shelley and increasingly unpopular with the British public, he started to exhibit symptoms of insanity. In a rare lucid interlude during his descent into madness, he remarked, "My mind, is, as it were, gone." In August of that year, he cut his throat with a penknife.

Curiously this wouldn't be the only suicide to cross FitzRoy's path. His eventual appointment as captain of the HMS *Beagle* and his association with Charles Darwin came about as the result of another suicide, this time of Captain Pringle Stokes, who had been carrying out a protracted hydrographic survey of the bleak waters near the tip of Tierra del Fuego onboard the HMS *Beagle*. "Nothing could be more dreary than the scene around us," Stokes wrote in his journal in 1828. He sank into a listless despondency fueled by the endless gray days and damp, bitter cold, and his diary entries became increasingly morose. While surveying the Golfo de Penas ("Gulf of Distress") on the Chilean side of the archipelago, he claimed it was a place where "the soul of man dies in him." A few weeks later, he locked himself in his quarters, refusing to come out. Six weeks after that, he shot himself in the head.

Enter Robert FitzRoy. At 22 years old, he found himself at the helm of what would turn out to be one of the most famous vessels in history. He proved an able captain and an excellent navigator; several years later, he circumnavigated the planet on a five-year voyage of exploration and discovery. FitzRoy asked Francis Beaufort, whom he knew well, to suggest an intellectual companion for that second voyage. The result was that a certain Charles Darwin was appointed chief science officer.

But this famous voyage was blighted. Over the course of the journey, FitzRoy's character revealed a venemously contrarian side that sometimes plunged Darwin and FitzRoy into fantastic arguments. Darwin referred to one of these in his diary as "bordering on insanity," and in a letter written years later he described FitzRoy as a "poor fellow, his mind . . . quite out of balance." FitzRoy, it seems, had little of the cool impartiality of the true scientist. He was one of the dying breed of Victorian amateur gentleman naturalists who would shortly be outstripped by professionals steeped in the protocols of logical methodology.

After the voyage, FitzRoy wrote and published a four-volume account of his adventures with Darwin. He received a gold medal from the Royal Geographical Society, and in 1841 successfully ran for parliament. In 1843, he was appointed governor of New Zealand. He had always been interested in meteorology and had remained in close contact with Francis Beaufort, who became a seminal partner in FitzRoy's aspirations to create a weather prediction system for seafaring vessels. FitzRoy returned to England in 1848 and in 1854 was appointed chief of a newly created meteorological department in the Board of Trade.

He began to standardize the collection of weather data from 15 inland observation stations in England, linked by telegraph to his office. In 1859, after a national maritime disaster, he seized on the opportunity to design weather charts for what he called "forecasting the weather." But his predictions were sometimes less than logical. In his 1863 publication, *The Weather Book: A Manual of Practical Meteorology*, he aired the peculiar notion that the length of the period between a sign indicating a change in the weather and the arrival of the weather in question indicated how long that period of weather would last. He strayed off the scientific path even more wildly when he had FitzRoy storm glasses installed at quayside in every major British port. These devices were to be consulted by sailors before they ventured out, but they were merely glass cylinders filled with a cocktail of potassium nitrate, ammonium chloride, ethanol, camphor and water. This mixture occasionally produced crystals or floating particles, and these, FitzRoy insisted, foretold changes in the weather. In truth, they had no connection to the weather at all.

Nonetheless, FitzRoy did have the distinction of publishing the first daily weather forecast in the *Times* of London. It would turn out to be his last hurrah. Times were changing, and quickly. The rapid transformation of the science of meteorology reflected how all the scientific disciplines were accelerating in the late nineteenth century. The epoch of the amateur was over. So it was up to FitzRoy's successor at the *Times*, Francis Galton (1822–1911), a Quaker, to create something that we would recognize today as a contemporary, scientific weather map.

PASSING THE TORCH

Galton was Charles Darwin's half cousin and a child prodigy. He was mathematically inclined and devoutly believed in the power of numbers. By 1861, he was the first to grasp the impact of the Coriolis effect on weather. Later the same year, he discovered that certain weather systems rotate in the opposite direction of cyclones and that, furthermore, the barometric pressure within these systems is higher than the surrounding region. He called them anticyclones. He then realized that cyclones and anticyclones were like cogs in a clockwork mechanism "and make the movements of the entire system correlative and harmonius." Galton began to improve on FitzRoy's weather maps, using lines (isobars) to link all the points of identical barometric pressure. These formed a series of concentric rings, a feature in almost every newspaper forecast today. The only thing his charts were missing, at least to contemporary eyes, were the graphics that represented weather fronts, an improvement still to come.

Galton brought considerable statistical weight to his postulations and published his results in the prestigious science journal *Nature*. His reputation began to eclipse FitzRoy's, who came to view Galton as his bitter nemesis. FitzRoy's health declined, and he fell into a severe depression. In 1865, when he was 59, FitzRoy, like his uncle before him, slit his throat.

Ten years later, on April Fools' Day, 1875, Galton's first weather map, showing detailed weather conditions across the British Isles from the day before, was published in the *Times*.

MATHEMATICS AND WEATHER PREDICTION

The science of meteorology entered the modern era with the help of four scientists who came after Francis Galton. They were an American, Cleveland Abbe; a Norwegian, Vilhelm Bjerknes; an Englishman, Lewis Fry Richardson; and a Hungarian, John von Neumann.

Cleveland Abbe was born in New York in 1838, six years before Samuel Morse sent the first telegraph message ("What hath God wrought"). The New York of Abbe's childhood would be unrecognizable today. There were no skyscrapers, no subways and no electric streetlights. The only electricity available was the minuscule current used by telegraphs. The winters were cold and the summers short and cool. The Little Ice Age didn't end until 1850, when Cleveland was 12. During a New York night, you could see the

Milky Way, and in the chilly mornings, on his way to school, Abbe would have heard roosters calling. It was downright rural.

Abbe was an autodidact at heart. When he was eight, his mother gave him a copy of William Smellie's *The Philosophy of Natural History*, a forerunner of the *Encyclopedia Britannica*, and he discovered a universe of information within its pages. Later, as a young man, he tried to enlist on the Union side of the Civil War, but he was rejected because of his myopia. He went to Harvard instead and upon graduation became a telegraph engineer, then worked as an astronomer at the Pulkovo Astronomical Observatory near Saint Petersburg, Russia, before assuming the directorship of the Cincinnati Observatory. It was here that his passion for meteorology flowered, and he started to map out an early warning system. He imagined that weather observers could be spaced out at regular distances, forming a grid of sorts, and they would be linked by telegraph to a central information processing headquarters. The Smithsonian Institution had already demonstrated the feasibility of such a scheme. By 1847, it was publishing telegraphically derived observations from weather watchers all over the United States, producing a weather map displayed daily in the institution's lobby. It became a tourist draw.

Abbe's dream of a national weather observation network would finally be realized in 1871 when he was appointed chief meteorologist at the National Weather Service. He enlisted 20 volunteer weather observers from across the country. At regular times, they were to transmit data on wind direction, temperature, precipitation and barometric pressure to his team of clerks at the National Weather Service. Another team would transfer the collected data to weather maps. For the first time, large weather systems could be followed and tracked hourly.

Abbe's data compilations and forecasts, "probabilities" as he called them, became a government service available to anyone with a telegraph. By 1900, the United States also had 114 automatic stations in operation along with its cadre of human observers.

At the height of his career, Abbe understood that the behavior of the atmosphere was the same as a fluid, writing that meteorology is "essentially the application of hydrodynamics and thermodynamics to the atmosphere." This insight turned out to be crucial, but the math necessary to elaborate his vision was beyond his capabilities. The probability torch was passed to the Norwegian scientist, Vilhelm Bjerknes.

CALCULUS OF THE CLOUDS

Bjerknes was a mathematical wunderkind born into a scientific dynasty. By age 15, he was a laboratory and research assistant to his father, Carl Anton Bjerknes, a physicist and the world's leading authority on hydrodynamics. Carl had developed a new resonance theory that drew parallels between the behavior of fluids and electromagnetism. Carl and Vilhelm were definitely not British aristocratic autodidacts with a penchant for natural history. Carl was chair of the pure mathematics department at the University of Oslo, and father and son were squarely in the vanguard of modern professional scientists.

Vilhelm Bjerknes received his masters of mathematics and physics from the University of Kristiania (later named Oslo) in 1888. He began to sketch out some of his most original scientific insights while still at the university, but in the increasingly competitive world of professional science, his father had developed a fear of publishing his own work, much less Vilhelm's. Painfully, Vilhelm realized he had to end his collaborative work with his father or be doomed to fall into obscurity. He set out on his own.

Vilhelm studied electromagnetics with Henri Poincaré in Paris, and then, in Germany, worked with the famous physicist Heinrich

Hertz. In 1895, he took up a professorship at the University of Stockholm and began to wrestle with the logistics of atmospheric hydrodynamics. He had married, and in 1897 his son Jacob, the next heir to the scientific dynasty, was born. Meanwhile, in a sort of Oedipal atonement, he began collecting his father's papers on hydrodynamics and published a two-volume set before his father's death in 1903.

In 1904, the year after his father Carl passed away, Vilhelm published a paper on how weather could be predicted using mathematical formulae. He devised a two-step forecasting rationale: diagnostic (what is), followed by prognostic (what will be). It was reminiscent of a grand theory that had been proposed a century earlier by Pierre-Simon, marquis de Laplace (1749–1827), one of the great mathematicians of the Age of Enlightenment.

Laplace had conjectured that science and mathematics would eventually converge into a system capable of predicting the future. In his treatise *Theorie Analytique des Probabilités*, written in the early decades of the nineteenth century, he proposed,

> *Given for one instant an intelligence which could comprehend all the forces by which nature is animated and the respective positions of the beings which compose it, if moreover this intelligence were vast enough to submit these data to analysis, it would embrace in the same formula both the movements of the largest bodies in the universe and those of the lightest atom; to it nothing would be uncertain, and the future as the past would be present to its eyes.*

Laplace was voicing the hubris of scientific rationalism at one of its most exciting periods. In this passage, he envisaged how a God-like intelligence might calculate the future given an omniscient knowledge of the present. It speaks to a profoundly Newtonian and Cartesian idealism, where elegant formulae paralleled the perfect harmonies of golden proportions, and it must have been at the back of Bjerknes's mind as he drew up his prognostic equations.

To enable his forecasts, Bjerknes devised a mathematical model that, in its basic form, is still in use today. His formulae were a phenomenal achievement, resting, as they did, on seven independant equations. Four dealt with each atmospheric variable — temperature, humidity, pressure and density (the continuity equation, the equation of state and the first and second law of thermodynamics) — while the other three consisted of hydrodynamic equations for motion. But there was a problem: he could not gather enough data about the initial state of the weather. No high-alititude instruments existed yet, nor did enough ocean-going ones. Furthermore, the equations were so complex that they could not be solved numerically or analytically. In theory, he could predict future weather patterns, but there weren't enough mathematicians to process the data.

HIGHS, LOWS AND WEATHER FRONTS

In 1917, while the First World War was still raging, Bjerknes founded the Bergen Geophysical Institute at the Bergen Museum in Norway. There he was assisted by his son Jacob, now 20 and a brilliant physicist in his own right. Jacob's friend Halvor Solberg and Swedish meteorologist Tor Bergeron joined them, and in less than a year, this extraordinary team had come up with the theory of "fronts," a term they adopted from the First World War battlefronts. They then went on to model what happens as a mid-latitude cyclone (low-pressure cell) is born, matures and then decays, and they introduced the symbols — semicircular red symbols for warm fronts and blue triangles for cold fronts — that we still use on forecast maps today.

Vilhelm Bjerknes's understanding of the atmosphere was a natural extension of Abbe's original insight: that, in essence, the atmosphere is a rarefied fluid. Like any liquid, it is liable to turbulence — the differential heating of night and day, summer and winter; the friction from mountains and the effects of landmasses and oceans — all of which creates eddies, like stirring milk in a cup of coffee. Francis Galton had divided these eddies into highs and lows, defined by their relative barometric pressure. It was well-known by Bjerknes's time that a high-pressure zone, or air mass, could either be warm or cold, as could a low-pressure air mass. Also well-known was the fact that warm air masses usually form in southern regions or over landmasses in summer, whereas cold-air masses form near the poles or over snow-covered landmasses in winter.

But regardless of their temperature, highs and lows have completely different origins. Bjerknes's team discovered that a high-pressure cell develops when air has time to accumulate in one region. This surplus of air grows cool and heavy at its lofty heights, and the chilled air flows downward and outward. Low-pressure cells are different creatures altogether, spawned by the collision of two high-pressure cells of different temperatures. Of course, highs and lows, aside from their pressure profile and genesis, differ in one very critical characteristic: their spin. The Coriolis effect causes a low to rotate counterclockwise and a high to rotate clockwise. In the southern hemisphere, the opposite holds true. It was the interaction and evolution of these cells that Bjerknes's team mapped out clearly for the first time.

A good example would be the birth of a low-pressure cell in North America. Here, lows often form when a warm, southern high-pressure zone bumps into a cool, northern high-pressure zone. Due to the fact that high-pressure systems rotate clockwise, the winds at the northern edge of the southern system blow in the opposite direction to the winds at the southern edge of the northern system. They are like cogs grinding against each other. The only "solution" to this problem is a swirl rotating in the opposite direction, counterclockwise. This is the beginning of the cyclone, which becomes a low-pressure area. As the air is drawn toward its center, the Coriolis effect sets it swirling in the opposite direction to the high-pressure region. Looking at the interaction of the two highs and the low, we cans see that the low acts like a lubricant between the two highs. It pulls an isthmus of warm air into and above the cool air mass. Now, as the southern high moves from west to east, a frontal system is created. The front is merely the leading edge, the boundary, between a mass of cold air and a mass of warm air.

Because highs and lows are generally circular, a front is usually curved, like the semicircles you see on weather maps. Their resemblance to military strategy maps was something that the Bjerknes's team didn't miss when they drew their first frontal system maps. But there's a lot more going on in three dimensions than a two-dimensional weather map reveals. Slicing through a front vertically, in cross section, you immediately notice that it is wedge shaped. Cool air sinks, hugs the ground, so that when a cold air mass is advancing and encounters a warm air mass, it wedges under the warmer air and pushes it up and over the cold front. The rising air carries water vapor through the dew point, which then condenses, first creating clouds and then rain. An advancing warm front is exactly the same, only in reverse — it pushes up and over cold air and squeezes it ahead of the front, creating almost the same wedge in cross section, although, as we'll see, with a difference. The low-pressure cell sits at the center of the action, mediating the contact zone between warm and cool.

To extend Bjerknes's battle metaphor, cold fronts might be called blitzkrieg attacks, and warm fronts fifth columns. Cold fronts have a steeper wedge and move faster than warm fronts, two reasons why their arrival is more abrupt. A cold front is always dramatic, creating thunderstorms in the summer and rain and snow in the winter. Warm fronts aren't as aggressive; they are more gradual, and their approach is easy to read. Because they move slowly and have longer, more tapered wedge profiles, sometimes hundreds of miles long, it's easier to predict a warm front's arrival. The sequence of clouds that ride the approach of the warm front begin with high cirrus that gradually transition into altostratus, which are then replaced by nimbostratus as the clouds lower and thicken over the retreating wedge of cold air. Rain or snow often accompanies the passage of a front because in both cases — either an advancing warm front or an advancing cold front — air rises, cools and creates adiabatic precipitation. The only fly in the ointment, at least in terms of accurate and timely meteorological forecasting, were those overly complex, time-consuming mathematical equations of Bjerknes's, which rendered practical meteorological calculations all but impossible.

But there was a new player about to arrive on the scene. Lewis Fry Richardson was another in the series of British Quaker polymath scientists. (The list of English scientists who were also Quakers is astonishing: the physicist Roger Penrose; his father, Lionel Penrose, the great geneticist; Luke Howard, who we've already met; Arthur Stanley Eddington, the world-renowned physicist; crystallographer Kathleen Lonsdale and Stephen Hawking's collaborator George Ellis, just to name a few.)

Richardson was familiar with Bjerknes's theories when, in 1913, he became director of a research laboratory for the United Kingdom's Meteorological Office. As a Quaker, he was a conscientious objector to Britain's involvement in the First World War, but he volunteered for the ambulance corps in 1916. (This was the same year that Cleveland Abbe was laid to rest, along with a copy of his beloved Smellie's *The Philosophy of Natural History*.) On furloughs between battles, Richardson worked on atmospheric equations as he sat on a bale of hay in the ambulance corps quarters. He took Bjerknes's calculations and revised them, replacing the finely graded analogs of calculus with measurements that sampled time in regular discrete portions, the way a strobe light breaks up movement into a series of stills. Each of these "moments" were approximations of change, but together, in sequence, they created accurate patterns. It was almost digital.

Now, he could do something no one had ever attempted: predict the weather mathematically. To do this, he needed a lot of meteorological information on a large landmass, and as it turned out the only weather charts detailed enough for his experiment were from the past. Here is an instance of "standing on the shoulders of giants," because Richardson went back to Bjerknes's very detailed weather charts for central Europe, 7 a.m., May 20, 1910.

He divided the map of the atmosphere into 25 modules, each measuring 125 square miles. These modules were further subdivided into five vertical layers of air. He plugged in all the variables and then ran the numbers that would, if his equations were correct, predict the state of the weather at 1 p.m., six hours later. It went disasterously wrong. He expected that the barometric pressure would be at 30.9 inches in six hours, but instead it remained "nearly steady." The defeat must have been stunning, but he went on to publish all his findings, including the failed forecast, in his 1922 publication *Weather Prediction by Numerical Process*.

Many decades later, his experiment was vindicated. Apparently the problem wasn't his math; it was the initial data, as Peter Lynch of the Irish Meteorological Service pointed out in 2006. Lynch repeated Richardson's experiment with the original data "initialized," as it would have been today, and the forecast became totally accurate. Thanks to Lynch, almost a century later, central Europe finally got an accurate forecast for the afternoon of May 20, 1910. Richardson had been afflicted with "scientific prematurity," a term coined by the fractal mathematician Benoit Mandelbrot years later.

And afflicted he was indeed. His theories were marginalized for decades after his 1922 publication, his self-confessed forecast failure having done little to promote scientific interest in his extraordinary set of forecasting equations. Despite the fact that he had simplified Bjerknes's math, his theorems still required an inordinate amount of calculations, and in the pre-computer era, it just wasn't feasible. "Perhaps some day in the dim future it will be possible to advance the computations faster than the weather adances," he wrote. "But that is a dream."

And yet his dream did come true, and probably sooner than he thought it would, thanks to John von Neumann.

Born into a Jewish Hungarian aristocratic family in 1903, John von Neumann was a numerical whiz kid. By age eight, he was proficient in differential calculus, and by 19 he had published two major mathematical papers. He accepted a professorship at the University of Berlin at the unheard of age of 23. His intellectual output was astonishing: on average he published a major mathematical paper a month. But being Jewish in the darkening climate of Europe held little future for him, and he knew it. When he was offered a professorship at Princeton in 1931, years before the scientific exodus of Jewish physicists from Europe, he took it.

By the mid-1930s, von Neumann had a reputation for tackling immense scientific and mathematical enigmas and became the go-to guy for technical problem-solving. He even rescued the first atomic bomb test from failure. The Manhattan Project was a rushed affair, and the bomb slated for the famous Trinity test had a subcritical mass of plutonium and was in danger of being a dud. Von Neuman used shaped charges, whose exacting contours he had mathematically determined, to implode the plutonium symmetrically to produce a successful chain reaction.

(Von Neumann's agile mind seemed to require stimulating environments. In the decades after the war, while working at the Princeton enclave, he often played German marching music very loudly as he worked on his theorems. So loudly, in fact, that a neighbor of his, a certain A. Einstein, complained that it was interfering with his concentration.)

Prior to his work on the atomic bomb, von Neumann was interested in hydrodynamical turbulence (the random eddies and currents within liquids) and the nonlinear equations that could describe it. These equations were essential to the ultimate mathematical descripton of the atmosphere, and von Neumann realized that some sort of electronic calculating device would be needed to undertake the vast computations necessary to solve the equations. So in 1946 at Princeton, he masterminded the construction of the world's first truly progammable electronic computer, the electronic numerical integrator and computer (ENIAC). Using the ENIAC, with the help of another genius mathematician by the name of Jule Charney, von Neumann programmed the first computer-generated forecasts in 1950. They were completely accurate. Charney sent the results to Richardson, then 69, in England. It must have been a wonderful moment for Richardson. His theories were vindicated, and today Richardson's numerical technique is the gold standard of weather forecasting.

A chapter in Richardson's 1922 book, *Weather Prediction by Numerical Process*, was devoted to turbulence, one of the most difficult phenomena to model mathematically. He acknowledged this complexity with a whimsical verse: "Big whirls have little whirls that feed on their velocity; and little whirls have lesser whirls, and so on, to viscosity . . ." If ever there was a literary equivalent to the Mandelbrot set, the famous fractal equation that creates a self-referential world of infinitely repeating patterns within patterns, then this was it. Richardson had intuited the forthcoming science of chaos and with it, a monumental challenge to forecasting.

THE BUTTERFLY EFFECT

"The weather is always doing something there; always attending strictly to business; always getting up new designs and trying them on people to see how they will go. But it gets through more business in Spring than in any other season.

In the Spring I have counted one hundred and thirty-six different kinds of weather inside of twenty-four hours."

MARK TWAIN ON NEW ENGLAND WEATHER

Edward Lorenz (1917–2008) had always been a weather buff. The changeable New England weather was sure to inspire a young meteorologist like Edward, and as a boy growing up in West Hartford, Connecticut, he built a small weather station in his parents' backyard, not unlike Luke Howard's some 50 years earlier. Its centerpiece was a special thermometer that kept an automatic record of the daily high and low temperatures with little sliding markers. He'd check the temperatures twice daily and record the numbers in a notebook. He also had a passion for math, although his twin interests seemed like two solitudes. He could measure the sensation of warmth and cold and translate those sensibilities into averages and means, but that's as far as it went. They weren't part of elegant, mathematical equations.

His mind craved logical challenges. On weekends and weeknights, he spent hours poring over problems in mathematical puzzle books, sometimes enlisting his father's help. As Lorenz got older, he began to lean more toward mathematics. In fact, after graduating from Dartmouth College in 1938, he went on to get a masters of mathematics from Harvard. But then the Second World War intervened.

The Army Air Corps needed meteorologists, and Lorenz was more than qualified with his Harvard degree, not to mention that he was still a passionate weather buff. He landed a plum posting as a military weather forecaster, a job that kept him out of combat. But the stakes were high. As the war deepened, the pressure to come up with accurate forecasts was fierce — other young men's lives were on the line. Richardon's equations were not yet in use, so accurate long-term forecasting was impossible. Meteorology was still an approximate science, based as much on intuition as it was on reading instruments or the look of clouds.

While Lorenz was second-guessing forecasts in the Army Air Corps, his fellow meteorologists were more interested in theory than pragmatics. The 1940s was a period when academic meterologists derided seat-of-the-pants forecasting. They much preferred the cleaner, more elegant theoretical side of meteorology, one in which potentially inaccurate forecasts didn't put their reputations at risk. But Edward gained a lot of hard experience during the war, and by its finish, he knew weather as well as any individual could. Yet he also had unfinished mathematical business. There was something about those seemingly random sequences of daily highs and lows that he had recorded as a child, something lurking behind the numbers.

Fifteen years after the war ended, Lorenz was on the faculty at one of the world's top research facilities, the Massachusetts Institute of Technology, and in 1962 he was appointed professor of meteorology. He had become a fixture at MIT with a reputation among his peers of being a little preoccupied and distant. That must have been a feat, given how many others in the faculty shared those characteristics. On top of which, he didn't look the part of a scientist — he had a down-home, somewhat rural look, a weathered face and piercing gaze.

It was Lorenz who first thought of using a computer to mimic the fluid dynamics of the atmosphere. In the late 1950s, at a time when small computers, especially ones that occupied less than a room, were hard to come by, the Royal typewriter company had released the Royal McBee, a "compact computer" the size of a large desk. It had a keyboard to enter programs and commands and a printer to output results. It was like a Macintosh computer decades before there were any, though it cost around $16,000, the price of a modest, two-bedroom bungalow.

Lorenz persuaded MIT to buy him one. (The rest of the faculty was skeptical about the ability of such a small computer to contribute anything meaningful to hard science, but they were all fascinated by his new toy.) He programmed it to simulate world weather patterns. In his Royal McBee was a microcosm of the world, a planet within the planet. He modeled prevailing winds, high- and low-pressure systems, temperatures — and then he let the whole thing run on its own. Every once in a while, he'd check up on what was happening by looking at printouts that translated the changing weather on this small, ideal planet into wavy lines on a graph.

One day while running the weather program, he decided to skip ahead and pick up the sequence in mid-stride, as it were. To set the machine at the initial conditions, he typed in the number from an earlier printout and let the Royal McBee run the equations again while he went out to grab a cup of coffee. When he returned, he knew something was wrong. The lines on the new printout were diverging from the original, even though the number he'd typed in was identical to the first sequence. Well, almost identical. He had shortened the sequence by a mathematically infinitesimal amount; it couldn't have produced such a divergence. Or could it? This is when Lorenz first intuited that weather sytems could be completely altered by the smallest change in the initial conditions. He rechecked his math and went to work. His revolutionary paper describing this process, delivered in 1972, was titled "Predictability: Does the Flap of a Butterfly's Wings in Brazil Set Off a Tornado in Texas?"

It was a bombshell, exploding in the midst of the increasingly confident field of computer meteorology. The early chaos scientists called the butterfly effect "sensitive dependence on initial conditions" wherein a small perturbation at the beginning could cascade upward through the whole system, altering everything. A popular folk saying sums it up:

For want of a nail the shoe was lost;
For want of a shoe the horse was lost;

For want of a horse the knight was lost;
For want of a knight the battle was lost;
For want of a battle the kingdom was lost!

It became a David and Goliath conflict. Lorenz's McBee had challenged the biggest computer in the world, the Cray supercomputer at the European Centre for Medium-Range Weather Forecasts in Reading, England, which used the von Neumann–Richardson algorithms. The stone in Lorenz's slingshot was his simple simulation program that successfully, as it turned out, modeled the susceptibility of the Earth's atmosphere to small initial changes.

If a supercomputer, with a far greater processing power than today's forecasting supercomputers, was connected to weather sensors stationed on every square mile of the Earth's surface and every square mile of the atmosphere and ocean depths, how accurately would it be able to predict the weather? According to Laplace's theory of the omniscient intelligence, if the machine were perfect, then it would match reality in lockstep for millennia. But according to Lorenz, and even some of today's meteorologists, it would fall behind reality in a relatively short period of time.

Jagadish Shukla, a climatologist at George Mason University, remarks that today's forecasting computers can forecast the weather fairly accurately five days into the future. But there's a limit, he insists, beyond which even the most powerful computer can venture: "We may not be able to get beyond day fifteen. [No] matter how many sensors you put in place, there will still be some errors in the initial conditions, and the models we use are not perfect . . . the limitations are not technological. They are the predictability of the system." Perhaps we will never, ultimately, be able to forecast much beyond a fortnight. We can know the climate; we can predict the average temperature of the ocean for decades, even centuries ahead; but we cannot know what will happen in Brazil or eventually in Texas.

9

APOLLO'S CHARIOT

THE SEASONS

The wolves are the first to sense it. For days now, they have been howling as they haunt the high Arctic darkness. It is late January in Grise Fiord, a settlement on the south shore of Ellesmere Island, and for three months the townsfolk have been locked in the icy talons of a pitch-black Arctic night. The wolf calls are a signal for everyone who lives here. The sun is coming. To the south, above the frozen ocean and low hills, an almost imperceptible daily brightening has begun. Week by week, this deep-red, noon-hour glow, a dawn rescinded daily, gets brighter and longer until it begins to rival the northern lights. Overhead, the Arctic stars have begun to dim.

At noon in the second week of February, something spectacular happens. The sun erupts above the horizon, For the first time in months, pure, unadulterated sunlight strikes skin, floods through windows and blinds eyes in an almost painful wound of light. This first sunshine, a day lasting only a few minutes, is spring in its purest form — raw and crystalline. The ice-hoary dwellings of Grise Fiord, like buildings long-submerged under the ocean and covered with coral encrustations, glow orange in the light, and every peak of the low hills surrounding the bay also glows with the same orange fire. Apollo's flaming chariot has arrived, and burning deep inside his golden brazier is summer itself.

Of course Apollo doesn't circle the Earth as the ancients contrived it. Instead, our planet circles Apollo; Apollo's brazier, as it happens, is 93 million miles away from us. The reason our trek around the sun is also our journey through the seasons is due to something else: an asymmetry, a quirk. The ancient Greeks knew this. They wove it into their mythology in a story where Zeus, Apollo's father, takes sides in the bizarre feud between two brothers, Atreus and Thyestes, who are contesting the kingship of Mycenae. Zeus favors Atreus over Thyestes, and when Thyestes captures a golden lamb, Zeus changes the course of the stars to express his displeasure. According to Roberto Calasso in his classic *The Marriage of Cadmus and Harmony*, this was nothing less than "an allusion to the tilting of the Earth's axis."

MIGRATING CLIMATES

So it's all in the tilt — spring, summer, fall and winter. As we rocket around the sun at 67,000 miles per hour, Earth's 23.26° pitch creates the seasons. If we weren't tipped relative to our orbital path (also known as the ecliptic plane), there would be no difference in the angle of sunlight hitting the Earth at different points of our orbit. There would be no midnight sun, no three-month polar nights, no solstices or equinoxes. Come to think about it, there would be no monsoons, hurricanes, typhoons or any great periodic winds either, because there wouldn't be a thermal imbalance between the parts of our globe that the sun seasonally favored or abandoned.

In a perfect Euclidean universe, all the planets in our solar system would rotate with their north-south axes at right angles to the ecliptic plane. But they don't. Since the birth of the solar system, some 4.6 billion years ago, the random impacts of large asteroids have skewed the rotational axes of every planet except one. So we're not the only tippy planet in the solar system. Mars has a 25.19° tilt, Saturn tilts 26.73° and Neptune has a tilt of 28.32°. The crazy planets are Venus, almost upside down with a 177.3° tilt, and Uranus, pivoted sideways at 97.77°. Only Mercury is nearly vertical. In terms of extraterrestrial seasons, it would seem that Mars is our closest celestial relative. Its thin atmosphere may be 97 percent carbon dioxide, but Mars has seasons like our own, even if each one lasts little more than five months, due to the Martian year being 687 days long. (Curiously, the length of a Martian day is pretty much the same as ours.)

Here on Earth, the number of seasons is dictated by latitude. The tropics have only two seasons: wet and dry. The subtropics have three seasons: a wet one, a dry one and a cool or mild one that can sometimes overlap with the other two. The classic four seasons (well, six actually) are only experienced where the winters are cold and the summers hot. These regions, the temperate zones, stretch from about 35° latitude (the average limit of the subtropical regions) all the way to the Arctic Circle in the northern hemisphere and the Antarctic Circle in the southern hemisphere. Los Angeles, Buenos Aires, Capetown, Osaka and Sydney, for example, are tucked just within the subtropical zone, whereas London, New York, Berlin, Moscow, Montreal and Beijing are well inside the temperate zone.

The six seasons in the temperate zone are hibernal (winter), prevernal (late winter, early spring), vernal (spring), aestival (summer), serotinal (late summer) and autumn. I like these nuanced divisions, particularly prevernal and serotinal, because they capture something of the magic of seasonal transitions. In the northern temperate zones, the prevernal marks the time when the sap begins to run in late February and when, in early March, a few patches of snow remain on the northern slopes of hills, but the grass is green on the southern sides. Here in North America, the serotinal season is that marvelous time, in early September, when the monarch butterflies start to migrate, the lakes are still warm enough for swimming and the resorts are empty. As Helen Hunt Jackson writes in her poem "September,"

By all these lovely tokens
September days are here,
With summer's best of weather
And autumn's best of cheer.

Seasons are migrating climate zones. I don't have to leave my house in Toronto to experience high Arctic or Amazon weather — they come to me. Here in the temperate zone, we import our climates, which arrive at crawling speed, roughly 0.7 miles per hour. Spring marches northward at about 17 miles a day, which means the first leaves open in London, England, 40 days later than in Barcelona, Spain. Leaf-opening in Winnipeg, Manitoba, lags behind Dallas, Texas, by a good 65 days. Summer recedes southward at the same speed as winter advances. Spring, summer, autumn and winter all recapitulate the ancient seasons of the Earth, and each is a temporal microcosm of the evolution of life itself.

When I was a child, it seemed life began on March 21, and even though it was often as cold as any day in February, there was an unmistakable intensity to the prevernal sunlight. Charles Dickens nailed it in *Great Expectations* when he wrote, "It was one of those March days when the sun shines hot and the wind blows cold: when it is summer in the light, and winter in the shade." By March 21, the sap was running in the maple trees. Sometimes my parents would tap our sugar maple in the backyard. I remember cold mornings where a thin cap of ice had formed over the sap in the bucket, and I would carefully lift the thin disc of ice out of the bucket and nibble bits off the edges, letting them melt on my tongue. The faint sweetness was ambrosial.

SPRING

"April is the cruelest month, breeding
Lilacs out of the dead land, mixing
Memory and desire, stirring
Dull roots with spring rain."

T.S. ELIOT, *THE WASTE LAND*

When I was growing up, there was usually a period of dry weather in early April, an arid spell when the first high-pressure cells of the spring established themselves over the warming land. Grass fires would inevitably start and the fire department would come to extinguish the blazes. I can still see the long, canvas hoses winding like giant pythons through the burnt grass beside the pond.

High over the April landscape, in the upper levels of the troposphere, the northward swell of heat pushes up on the boundary between the troposphere and the stratosphere. The troposphere above the temperate zone, usually 11 miles high, gains an extra mile in height by midsummer. It's as if the atmospheric infrastructure of the tropics expands its territory and, as the seasons migrate northward into the northern hemisphere, it seems anything with wings migrates with them. Sometimes insects will hitch a ride on a warm high-pressure system as they did in April 2012, when thousands of red admiral butterflies flooded southwestern Ontario, riding a northbound front up from Texas. One windy afternoon, my backyard was suddenly alive with these black, red and white butterflies.

Birds are famous migrators, often traveling thousands of miles, and their noisy return to their summer breeding grounds marks the official beginning of high spring. The famous swallows of San Juan Capistrano overwinter in Argentina and arrive in the southern Californian town almost always on the vernal equinox itself, March 21. Pretty punctual after a 6,000-mile flight.

The summery high-pressure cells that begin to preside over the northern temperate zones in May are not shared by the subtropics. There, the Earth's tilt pushes the intertropical convergence zone — a rainy, global belt of low pressure caused by converging trade winds — northward. By June, the convergence zone shows up in the Caribbean as the rainy season. In the southern hemisphere, the opposite holds true, and the intertropical convergence zone slides south of the tropics in December.

By late spring in the temperate zones, the brand-new foliage exhales humidity and, in the sunlight after a summer shower, mists rise. The rainforest climate has moved in. Storms also ride north with the hot weather of late June, as thermals build cumulonimbus clouds in the humid air. The lilacs have finished blooming and, in my hometown, by the end of June the shallows at the edge of the pond are inky with tadpoles, some of which are already growing legs in a recapitulation of the evolution of fish to amphibians. The June peonies explode into blossom all at once. As Michael Cunningham wrote in *The Hours*, "What a thrill, what a shock, to be alive on a morning in June."

INVINCIBLE SUMMER

"Summer afternoon — summer afternoon; to me those have always been the two most beautiful words in the English language."

HENRY JAMES

Every winter I yearned for it, every spring I waited for it and every June, like a dependable miracle, it arrived. Summer was, and still is, the universal panacea, an antidote to adult worries, a time, wrote Ada Louise Huxtable, "when one sheds one's tensions with one's clothes, and the right kind of day is jeweled balm for the battered spirit. A few of those days and you can become drunk with the belief that all's right with the world." Even the misery of poverty is somewhat remedied by summer, when "the living is easy." For children especially, summer in the temperate zones is the gateway into an enchanted territory of imagination. As Lewis Carroll wrote in "Solitude,"

I'd give all the wealth that years have piled,
The slow result of Life's decay,
To be once more a little child
For one bright summer day.

Of the four major seasons, none is as popular or as beloved as summer. It is then that we reestablish our sensual contract with nature. The beginning of summer in the northern hemisphere, the solstice on June 21, is not just six months away from the winter solstice in December, it is also 186 million miles away as the space crow flies. From solstice to solstice, the Earth has traveled half of its 584-million-mile journey around the sun.

In late spring and early summer, the jet stream migrates north along with the northward edge of the Ferrel cell, pushing the cooler polar air into the northern reaches of the temperate zone. With the arrival of summer, our senses are reawakened — the smell of roses, the hissing of lawn sprinklers, the sound of surf and children laughing, the wind in the leaves, the drone of lawnmowers. The intoxicating, silky swish of a summer breeze against bare skin. Ice cream and figs. The margin between indoors and outdoors is blurred, and the summer landscape turns into a vast, shared living room. With a pair of folding chairs and a picnic table, a backyard becomes a dining room and the local park becomes a stage for *A Midsummer Night's Dream*. As Henry David Thoreau wrote in *A Week on the Concord and Merrimack Rivers*,

> *In summer we live out of doors and have only impulses and feelings, which are all for action . . . we are sensible that behind the rustling leaves, and the stacks of grain, and the bare clusters of the grape, there is the field of a wholly new life, which no man has lived; that even this earth was made for more mysterious and nobler inhabitants than men and women.*

HEAT LAGS AND DOG DAYS

Water warms more slowly than land, but even land takes time to absorb solar heat. This means that although the longest day of the year is June 21, it isn't until the last week of July that summer heat climaxes. (A week later marks the peak temperature of North America's southern Great Lakes — 21°C). In western Europe, the heat lag is shorter, more like a month, so the hottest days are in July. In San Francisco, because of the cold Pacific gyre, the summer lag is almost two months. Their warmest weather peaks in mid-August to September.

Chronologically speaking, the exact middle of summer falls on August 6, halfway between the June solstice and the September equinox. In Italy, they celebrate the height of summer nine days later, on August 15, with Ferragosto, a holiday that has a 2,000-year pedigree. In North America, the spell of hot, still weather around this time is referred to as the "dog days" of summer, a phrase you'd think originated from the behavior of dogs during heat waves: they lie down and pant. Actually, the phrase comes from the ancient Greeks and refers to the 40 days of summer, from July 3 to August

11, when the dog star Sirius rises above the southern horizon. This is also often the period of least rainfall in the northern hemisphere due to stable high-pressure cells setting up over temperate zones.

During heat waves, especially under a jet stream pattern locked in place, a temperature inversion can form and make things even hotter. It's a reversal of the usual adiabatic trend of the atmosphere to get cooler with altitude. Instead, a layer of warm air forms at an altitude of 2,000 to 3,000 feet and traps relatively cooler air beneath, sometimes for weeks on end. I say "relatively cooler," because during a heat wave, the temperature of the first thousand feet of the atmosphere can get very hot indeed. One sweltering July in 1995, Chicago, Illinois, suffered through a weeklong heat wave caused by an inversion layer. On the 15th of that month, the city recorded a daytime high of 37°C. The ERs were flooded with patients suffering from heat stroke and dehydration. But as heat waves go, that was mild.

The world's longest recorded heat wave struck down under, in western Australia. The citizens of Marble Bar watched their thermometers climb to 38°C every day for five months, starting in November 1923 and lasting until early April, 1924. (It must have been one scorcher of a Christmas.) The hottest day on record wasn't in Australia though; it was in Kuwait, in a small town called Mitribah. There, on July 21, 2016, the daytime high reached a searing 54°C.

People have always complained, "It's not the heat, it's the humidity," although no one had accurately measured that subjective experience until it was quantified by two Canadian meteorologists in 1979. J.M. Masterton and F.A. Richardson took the dew point as the index of humidity relative to the temperature and produced a scale that was more accurate than the American heat index system, which is based on relative humidity alone. Their humidex reading is a single temperature, in Celsius or Fahrenheit, which accurately reflects how our bodies sense the heat. For example, if the temperature is a sultry 30°C and the dew point is a relatively dry 15°C, then the humidex number is close to the actual temperature at 34°C. But if the dew point is raised to 25°C, then the humidex number is a whopping 42°C.

Often the only relief from daytime heat is the cooling that night brings. As Henry Wadsworth Longfellow observed in his notes, "Oh! how beautiful is the summer night, which is not night, but a sunless yet unclouded day, descending upon earth with dews and shadows and refreshing coolness." In clear weather, even during the most extreme heat wave, the temperature starts to fall at sunset and continues to fall overnight until it reaches its lowest point just before dawn, usually a drop of 7°C. If there are clouds at night, then the cloud layer traps the heat and the temperature drops more slowly, or not at all. Which is why people like to spend their dog days at the lake or by an ocean.

Even in the absence of prevailing winds, a large body of water will set up its own cycle of winds during hot weather. If the land is hotter than the water, there will be an onshore breeze as the updraft over the land draws in cooler air from the lake or ocean. Conversely, if the body of water is warmer, then an offshore breeze will set up at night as cooler air from the land is drawn into the updraft from the water. That's one reason why people vacation by lakes and oceans. And by mountains.

Dry air cools adiabatically with height, so there's immediate relief from the heat the higher you go. Mountains also amplify any prevailing wind the same way that tall buildings funnel air around them to blow your umbrella inside out. As well, even the stillest summer night can be interrupted by another kind of breeze altogether, a surprisingly strong, cool wind that flows down mountainsides. These katabatic winds are the result of cool air collecting on the mountain heights. Because cold air is heavier than warm air, the air begins to flow down the slopes like water off a roof, gaining momentum, not to mention heating adiabatically, as it descends. On a big mountain, these flows can become gale force winds, cool on the inclines and warm in the valleys. I'd like to imagine that is where Lucy Maud Montgomery's lovers met in her novel *The Blue Castle*: "It was rapture enough just to sit there beside him in silence, alone in the summer night in the white splendour of moonshine, with the wind blowing down on them out of the pine woods."

AUTUMN

"'Tis the last rose of summer,
Left blooming alone;
All her lovely companions
Are faded and gone."

THOMAS MOORE

There is something exhilarating about the first cool days of autumn in the northern temperate zone. It's not just because it's back to school with new shoes and colored pencils; it's because we are mammals. Autumn is our native season. We came into our own in the cool forests and savannahs of the Pliocene-Quaternary glacial period. In fact, all the temperate mammals underwent a final spurt of evolution at about the same time — dogs, cats, squirrels, chipmunks, hedgehogs, porcupines, bears, deer. And now, as the nights get cooler, they grow new fur coats to get them ready for winter. Change is coming, and every one of us mammals feels it in our veins.

Autumn has a definite fin de siècle ambience, but it also holds urgency; the preparation for the deep freeze gives it a sense of purpose. The chill air quickens the pace, and for many this is their favorite season. In "The Autumnal," the poet John Donne insists, "No spring, nor summer beauty hath such grace, / As I have seen in one autumnal face." And Percy Bysshe Shelley was definitely a believer. In "Hymn to Intellectual Beauty," he wrote, "There is a harmony / In autumn, and a lustre in its sky, / Which through the summer is not heard or seen, / As if it could not be, as if it had not been!"

With the warm, serotinal weather continuing in September, the deserted beaches become a poignantly vacant stage for summer's last drama, its decline. Arthur Symons's poem "At Dieppe" (written in 1895, when Dieppe was a holiday destination, not a Second World War landing beach) catches the deserted-resort mood perfectly: "The gray-green stretch of sandy grasses, / Indefinitely desolate; / A sea of lead, a sky of slate; / Already autumn in the air, alas!" The same mood pervades Matthew Arnold's poem "Rugby Chapel,"

> *Coldly, sadly descends*
> *The autumn evening. The field*
> *Strewn with its dank yellow drifts*
> *Of withered leaves, and the elms,*
> *Fade into dimness apace,*
> *Silent; hardly a shout*
> *From a few boys late at their play!*

There is a pensive, meditative distance in these poems. But the atmospheric machinery of autumn is anything but languorous. As the northern hemisphere's polar cell enlarges its frigid kingdom at the expense of the temperate mid-latitude Ferrel cell, the jet stream begins its 800-mile slump southward. By mid-autumn, it is already 400 miles south of its summer position, and northwest breezes are bringing the first cold nights. The leaves begin to turn, and, as fall progresses, their color deepens until the polychromatic autumn foliage takes on an almost hallucinatory ambience. In his book *Songs: Set Two — A Short Count*, the American poet Ed Dorn refers to the fall colors he witnessed once when crossing the border from upstate New York into Canada: "We done the crossing of the border / In the lysergium of September."

I also feel the mammalian tug, the thrill of autumn, though it is outweighed by tremendous nostalgia for my favorite season, summer. For the poetic soul, the autumn can instill a vast, pathetic fallacy of lost paradise and poignancy, a mourning of departed love. It is unclear, for instance, if Alfred, Lord Tennyson was writing about summer or a lost love — or both — when he wrote,

Tears, idle tears, I know not what they mean,
Tears from the depth of some divine despair
Rise in the heart, and gather to the eyes,
In looking on the happy autumn fields,
And thinking of the days that are no more.

The same ennui infuses Paul Verlaine's poem "Chanson d'automne": "The long sobs / Of the violins / Of autumn / Pierce my heart / With monotonous languor."

There is an even subtler reading of autumn, one that invokes the slightly spooky, somnambulistic ambience of late October and early November, when the short days yield to the somber empire of twilight. Ray Bradbury evokes the mood precisely in his short story "The October Country":

That country where it is always turning late in the year. That country where the hills are fog and the rivers are mist; where noons go quickly, dusks and twilights linger, and midnights stay. That country composed in the main of cellars, sub-cellars, coal-bins, closets, attics, and pantries faced away from the sun. That country whose people are autumn people, thinking only autumn thoughts. Whose people passing at night on the empty walks sound like rain.

I think that the height of poignancy is not reached in September, during long walks on deserted beaches or passing by an abandoned amusement park on a rainy afternoon, but later, during "Indian summer" — summer's aftershock, that last blast of warm weather in late October. In Europe, it's called St. Luke's summer or St. Martin's summer, heralding a period of fine, warm weather that falls between October 18 and November 11. It is summer's last hurrah.

Although fall officially lasts from September 21 to December 21 in the northern hemisphere (March 21 to June 21 in the southern hemisphere), winter's invasion can often start with an overnight dusting of snow in October or November. Winter ends the seasonal cycle of evolution. It is the great null space that resets the clock to zero, just as it did in prehistory 700 million years ago, when a glacial age nearly eradicated life on our planet.

When I was young, during the first snowfalls of late autumn, I imagined herds of wooly mammoths gathering in the frosty gloom at the foot of glaciers, blue-green cliffs of ice that towered a mile high. And as the great mountains of ice ground their way southward, they crushed everything beneath them into powder and slush.

10
A COLD PLACE
WINTER AND THE ICE AGES

"It was so cold I almost got married."
SHELLEY WINTERS

A few days after Christmas, on a whim, you accept an invitation from a friend to celebrate New Year's Eve at his cottage. When you board the bus in the early afternoon, it's sunny and cold. The forecast calls for a high of -10°C with a light southwest wind, though a polar front is predicted to move in that evening. An hour later, as the bus reaches the turnoff from the highway, the wind picks up, and it's beginning to snow, probably lake-effect snow flurries. You realize that the polar high must be moving in faster than predicted. Within another hour, it's snowing heavily, and your bus seems to be the only vehicle on the road. You walk to the front and ask the driver how much longer to your transfer. "Fifteen minutes," he says, and then adds, "There's a weather advisory. I hope you've got something warmer than that coat. Those shelters can get pretty cold this time of year." You tell him you don't expect you'll be waiting very long — the connecting bus is scheduled to arrive less than half an hour after you get dropped off.

At the transfer, the snow is so heavy you can barely make out the shelter in its headlights. After the bus leaves, it is surprisingly quiet — the silence of heavy snow. The shelter, built to resemble a cheery log cabin with three walls and a roof, is empty of passengers. You sit down to wait. It's really cold. You take out your phone but there's no service. You're beginning to regret not wearing a heavier coat. At least you have a scarf and gloves. After an hour, you begin to wonder if the bus is coming. No cars are passing by. Have the police closed the highway? Night is closing in. Another hour passes, and you decide to start walking to get somewhere warm, back down the road. Wasn't there a gas station near the turnoff?

You take a sweater out of your pack and pull it on along with an extra pair of socks. Setting off down the road into the accumulating snow, which, you notice with some alarm, is already up to your ankles, you realize you're not dressed at all properly for this deep freeze — your coat is too light and your gloves are thin leather. You should have brought a hat. At least you're wearing winter boots. You calculate how long it will take to walk back to the turnoff from the highway. Three, maybe four hours. The wind on your face is so cold it burns. With the arrival of the polar front, the temperature has already fallen below -15°C. Adding in the wind chill, it feels more like -22°C.

You plod on into the blowing snow. Sometimes you're not sure where the road is, and twice you slip and fall. Daylight fades to night. Your phone's flashlight only illuminates a dense cone of falling snow. You can see better when you switch it off. You try checking the time, but you're shaking too hard. You're well into the first stages of hypothermia.

Extreme cold is much more lethal than extreme heat. The British medical journal *The Lancet* analyzed the cause of death data from 13 northern and temperate countries between the years 1985 and 2012 and found that 5.4 million were from hypothermia. A higher percentage of these deaths were men, because men are more susceptible to hypothermia than women. But the physiological process follows the same course for both sexes.

Without a hat, you're losing 10 percent of your body heat through your head. The snow has been building up in your hair, turning into a dripping helmet of ice that trickles down your neck. After an hour and a half outside, improperly dressed as you are, your core body temperature has fallen to 35°C, which is why you are shivering uncontrollably and violently. Your muscles are now cooling and tightening, causing you to walk in a stiff-legged, lurching gait.

You recall a passage from a book on meditation you read years earlier, about a monk who was tested by his master in the middle of winter to see how well he could control his metabolism. On a bitterly cold night, the master chipped a hole in the ice of a frozen river, and the naked acolyte sat beside the hole in the lotus position. The master dipped a blanket in the water and draped it over the shoulders of the acolyte who, over the next hour or so, had to dry the sheet with body heat. The master dipped the sheet in the icy water once more and the whole process was repeated until the acolyte had dried the blanket three or four times.

The reminiscence makes you feel chillier still. Your hands, even though you're holding them under your armpits, throb painfully. They have cooled to 18°C. The outside temperature continues to drop as the polar front establishes itself. It is -20°C; with the wind chill factored in, it's -27°C. At -28°C, exposed skin will freeze within 10 minutes. The tip of your nose is already numb, but something more insidious is happening. Now, for every degree that your core temperature drops below 35°C, your cerebral metabolism rate drops 4 percent. Your thought processes are becoming irrational, and you are starting to suffer brief periods of amnesia. You are very cold, very tired. You have no idea how far you've come when you realize what a fool you've been. Why not let the insulating properties of the snow work for you. You're too tired to keep walking anyway, and it's too far to walk back to the bus shelter. Besides, you can hardly keep standing.

At the side of the road, you scoop out a hollow in snow and lie down, fluffing some snow over your body. Your core temperature is 32.2°C. Despite your snowy duvet, it's falling a full degree every 30 minutes. When you reach 31°C, the shivering stops. A few minutes later, you have an overpowering urge to urinate. Your kidneys are being swamped by the overflow of fluid that the cold is constricting out of your extremities. This is the second last thing you'll remember. At 30°C, your heart rate becomes irregular and slows. You begin to hallucinate voices, city sounds, flashes of light. These are accompanied by a sensation of growing warmth. Fortunately, you lose consciousness and don't experience the paradoxical sensation of burning fever that often overcomes those whose core temperature falls below 29°C, which is why frozen hypothermia victims are sometimes discovered with few or no clothes on. They've torn them off in an ironic, and fatal, frenzy.

You're lucky. You regain consciousness in a hospital room. Two teenagers on snowmobiles discovered you by pure serendipity when the vapor plume from one of your very occasional exhalations was caught in their headlights. By that time you'd been completely buried in snow except for your small exhalation vent. Fortunately, the emergency doctor who treated you knew what to do. Cutting into your abdominal cavity, she inserted two catheters, one to inject warm saline solution and another to drain it after the warmth had irrigated your stomach and intestines. Gradually, very gradually, she increased the flow of warm fluid, knowing that if she raised your temperature too quickly, you'd go into cardiac arrest and die.

So you survive. But you've been to a very cold place, a place that most people don't experience. You've taken the absolute cold into your body, and the mark of winter will be left at the tip of your nose, which you will lose to frostbite, along with the tip of one of your fingers. You have experienced what few of us really understand: that we exist in an illusion of warmth. No matter where we live on this planet, in the steaming equatorial rainforests of the Amazon or the wincing heat of the Sahara, a bone-chilling temperature of -20°C is never further than 26,000 feet away — straight up. And up just gets colder. The average temperature of interstellar space is -270°C. Warmth is the exception in this universe.

THE COLDNESS OF WINTER

"Behold the frostwork on the pane — the wild, fantastic limnings and etchings! Can there be any doubt but this subtle agent has been here! Where is it not? It is the life of the crystal, the architect of the flake, the fire of the frost, the soul of the sunbeam. This crisp winter air is full of it. When I come in at night after an all-day tramp I am charged like a Leydon jar; my hair crackles and snaps beneath the comb like a cat's back, and a strange, new glow diffuses itself through my system."

JOHN BURROUGHS

"In the bleak midwinter
Frosty wind made moan,
Earth stood hard as iron,
Water like a stone;
Snow had fallen,
Snow on snow,

Snow on snow,
In the bleak midwinter,
Long ago."

CHRISTINA ROSSETTI

More than a decade ago, I visited Tobago, an island in the Caribbean just off the coast of Venezuela. There, in the coastal village of Bucco, I befriended a local skipper, a fisherman by the name of David, who piloted the glass-bottom boat out to the reef. He was fascinated by my stories about Canada and he seemed particularly interested in my descriptions of winter, about the intensity of the cold. By way of personal experience, he recounted how he had been at the top of a high hill in Tobago one evening and it was so cold he saw his breath. Was that like the winter in Canada?

I told him that seeing your breath was just the beginning. "You know the freezer compartment in a refrigerator?" I said. "That's winter. That's what it's like outdoors in January. It's so cold that everything is frozen rock hard, the earth, the lakes and the leafless trees." I could see he was impressed, though there was a slight sense of incredulity to his amazement; after all, he was a reasonable person. How did people survive in such an extreme climate, and why on earth would they build cities that far north?

Sitting on the beach in Tobago, you'd never guess that we were in the middle of an ice age, but we are. If we weren't, no snow would fall anywhere, not even in the Arctic. Any period of Earth's history where ice covers one of the poles is considered an ice age, and the current one is called the Pliocene-Quaternary glaciation. Right now, massive glaciers cover Greenland, most of the Arctic islands and the entire continent of Antarctica. They also lurk at the summits of many mountains, waiting to pounce. Indeed, one large mountain glacier has the audacity to dwell on the equator, atop Chimborazo in Equador.

It will stay put, probably for thousands of years, because our ice age, which began less than four million years ago, is characterized by alternating periods of cooler and warmer climates. The colder periods, when the continental glaciers expand, are called stadials, and the relatively warm periods, when the glaciers retreat, are called interstadials. Currently we are enjoying a moderately warm interstadial period, although it's nowhere near as warm as the last one, the Sangamonian interglacial, which peaked about 120,000 years ago, around the time when our species, *Homo sapiens*, began to expand north from South Africa.

But if we got started during a warm interstadial period, we came of age during a major glaciation event, the Wisconsin glaciation. Winter shaped our species. Our versatile and playful minds grasped the frictionless potential of ice and snow, and we used it. Long before the wheel was invented, snow sleds were common in northern Europe. A flat board, or two runners, placed under a load made it much easier to drag over the snow. It is likely that Cro-Magnons wore something like skis to to pursue game, and fragments of wooden skis dating back to 6,000 BCE have been discovered in northern Russia. But likely even earlier than that, in eastern North America, the indigenous people developed my favorite winter sport — snowsnakes — which is played to this day.

In the coldest week of February, the small town of Brantford in southern Ontario plays host to a gathering of the Six Nations: the Mohawk from upstate New York, the Cayuga, Tuscarora, Onondaga and Oneida from Ontario, the Seneca from Michigan as well as the Delaware from Ohio. There they compete in the annual winter tournament of snowsnakes. The snakes themselves are exquisitely honed and lacquered wooden lances, some six feet long, whose snake-like heads are inlaid with finely wrought intaglios of metal. These lances are hurled with professional skill onto a narrow luge track that stretches more than a mile and a half. Whoever's snowsnake goes the farthest is the winner.

Speed is winter's gift. I can remember February weekends when all the kids on my street were tobogganing down the highest hill in the neighborhood. If they didn't have a sled or a toboggan, a piece of cardboard would do. We also had a frozen pond for hockey games. I used to skate all afternoon there, until my feet were numb. I was continuing a tradition, it turns out, thousands of years old. Five-thousand-year-old bone skates have been uncovered by archeologists in Scandinavia and Russia. Four and a half millennia later, just in time for the Little Ice Age, the Dutch invented iron skates, similar to the ones we use today.

Previously, during the medieval warm period, which lasted almost 700 years from 800 to 1450 CE, the climate was so balmy that Vikings established farms in Greenland, and England hosted extensive vineyards. Geoffrey Chaucer, who lived from 1343 to 1400, was a product of the last century of the medieval warm period. His grandfather and his father were wealthy vintners who could afford Geoffrey's education and an entrée into royal court society. The Norse farms and English vineyards ended with the arrival of the Little Ice Age, which lasted 400 years from 1450 to 1850.

Climatologists have a time machine of sorts, one that allows them to see what winters looked like back then. Pieter Bruegel the Elder's painting *The Hunters in the Snow* (1565) is one such marvelous window into the period. Two Dutch hunters and a pack of dogs plod through deep snow on a moody, overcast winter day in Holland. Behind them, in the distance, frozen ponds are covered with diminutive skating figures. This scene would be impossible in today's Netherlands. Hans Neuberger, former head of the department of meteorology at Pennsylvania State University analyzed *The Hunters in the Snow* and approximately 12,000 other paintings that dated between 1400 and 1967 from European and American museums. Measuring the percentage of cloudiness and darkness in paintings of the outdoors, Neuberger tracked the progress of the Little Ice Age and discovered a peak between 1600 and 1649 that almost exactly coincides with the beginning of the Maunder minimum (1645 to 1710), with the coldest period spanning the 40 years from 1670 to 1710. It is thought that this protracted cold period was caused by a dimming of solar activity characterized by a complete lack of sunspots; hence "minimum." The eponym "Maunder" is a tribute to solar astronomers Annie Russell Maunder (1868–1947) and her husband, E. Walter Maunder (1851–1928), who were the first to link solar activity to climate change. During the 40 years of the Maunder miniumum, the Thames River froze solidly enough that for decades winter fairs were held regularly on the thick ice.

Toward the end of the Little Ice Age, it was still cold enough to ensure that 1816 was the "year without a summer," in the United States. New England experienced heavy snowfalls in June, July and August. In Savannah, Georgia, that same year, the Independence Day high was only 8°C. (Of course the concurrent eruption of Mount Tambora in Indonesia didn't help matters that year.) Yet none of these Little Ice Age minimum temperatures compare with Fairbanks, Alaska, in January, where the average daytime high temperature is -18°C, or to one frigid day in February 1947, when the inhabitants of Snag, Yukon, watched their temperature drop to -63°C! But there's no competition when it comes to the coldest place on Earth — the home turf of the Pleistocene glacial age.

ANTARCTICA

"The squeak of the snow will the temperature show."

WEATHER PROVERB

Antarctica is the heart of winter, with temperatures rivaling the surface of Mars and the entire continent scoured by inhumanly cold winds. Explorers trying to reach the South Pole on foot were the first to battle the dreaded subfreezing katabatic winds, caused by cold air flowing down the continental gradient toward the ocean, gaining speed as they move. Gusting at a more-or-less constant 40 miles per hour, they sometimes pick up to hurricane-strength blasts of 200 miles per hour. Scott's ill-fated expedition to the South Pole was wiped out by these winds in 1912.

The coldest temperature measured on our planet since we have been keeping records was in Vostok, Antarctica, on July 23, 1983, when the mercury dropped to -89.2°C. Vostock experienced some of its warmest weather during the summer of 2003, when temperatures reached a balmy -32°C. By comparison, among the temperatures that a Viking lander recorded on Mars, the highest was -17.2°C while the lowest was -107°C. So, although the long winter night in Vostok does not rival the deep freeze on Mars, Vostok's warmest weather can be colder than a Martian summer. Which is not to say that temperatures on Earth have never fallen as low as those on Mars.

SNOWBALL EARTH

Imagine you are our time traveler again, this time visiting our planet 600 million years ago. When the time mist clears, you step out of your machine into the middle of a polar landscape, a vast, snowy plain that extends from horizon to horizon. It's fortunate you packed an Antarctic expedition suit because, even though the sky is a cloudless blue, the outside temperature is -35°C. You scrape at the crust of wind-hardened snow beneath your boots to reveal the icy surface of a frozen ocean. If you drilled through that layer of ice, you'd have to bore down 65 feet before you hit salt water. But you don't have time for that.

Even through your thermal insulation, you're beginning to feel the chill. Equally disturbing is the fact that the sun, directly overhead, tells you that you are in the tropics, possibly standing on the equator itself. How could that be? Welcome to Snowball Earth.

SPITSBERGEN'S SECRET

W. Brian Harland was a mid-twentieth century geologist from Cambridge University who specialized in glacial physiography. He had an encyclopedic memory and a reputation for being tenacious, cantankerous and a dogged researcher. He was also a Quaker, yet another in a long line of Quaker scientists. He liked to do his geology in the field, and the Norwegian island of Spitsbergen became his real-world geological laboratory.

From 1938 to 1981, he made dozens of field trips to the island and the surrounding Svalbard archipelago to observe, among other things, "galloping glaciers," ice-gouged fjords, towering moraines and meltwater sediment. He also explored and collected samples from the bedrock beneath the glacial deposits, and there, in the sedimentary rocks, Harland uncovered a mystery.

Spitsbergen was primarily composed of strata that had been laid down in shallow, tropical seas during the Devonian period. But on the west and north coasts, Harland found much older rocks, sediments laid down in the Precambrian-era tropics. Within these, he discovered tillites, the telltale deposits from glaciers. Problem was, the rock was 600 million years old — much, much older than any known glacial deposits. If these glacial tillites had indeed been formed in the tropics, what were they doing there?

Harland experienced his first intimation of a monstrous, heretical idea. If glaciers had extended all the way to the tropics, then the whole planet must have been frozen. Perhaps there had been a catastrophic, global ice age, one so extreme it made recent glacial ages look like spring frosts. But he needed more evidence, so between field trips to Spitsbergen, he began a search through geological collections from all over the world. He was looking for 600-million-year-old rocks. He was particularly interested in dropstones.

As a glacier scrapes over bedrock, it picks up stone fragments and carries them along. If a glacier melts on land, it leaves piles of ice-rounded stones in drifts called moraines. But if a glacier extends over water, the stones fall beneath the floating glacier to the bottom of the lake or ocean. These are dropstones. And Harland found them everywhere, in every collection of 600-million-year-old rocks from around the world.

His hunch was right: there *had* been a worldwide ice age. He published his findings in 1963 and braced himself for scientific fame. Instead, his geological colleagues dismissed his theory as ludicrous. The tropics could never freeze, they argued, not now, not 600 million years ago. They insisted that the rocks that Harland had analyzed were not in the tropics when the dropstone layer was formed. Continental drift had carried them far from where they had originally been deposited. Harland's global deep freeze theory seemed to have come to a dead end. He had to wait more than 20 years before it was resurrected.

MORE THAN A CHANCE IN HELL

In the late 1980s, Joseph Kirschvink at the California Institute of Technology became fascinated by Harland's theory. As the world's foremost expert in geomagnetism, Kirschvink was in a unique position to prove or disprove Harland's global ice age hypothesis. He knew that when a rock is formed, it is permanently imprinted with the direction of the Earth's magnetic field lines. Rocks that are formed in the tropics have horizontal field lines and those from the Arctic have almost vertical ones. Kirschvink had just constructed an extremely sensitive magnetic field detector in his laboratory, and he decided to put Harland's theory to the ultimate test. He was prepared for the worst. "Many wonderful, beautiful theories," he said, "have been slaughtered by a nasty little fact."

So, like Harland before him, Kirschvink acquired samples of 600-million-year-old dropstone strata from all over the world. He then analyzed them in his geomagnetic detector. What he found astonished him. Many of the rocks *had* been formed in the tropics. Harland's vision of a planet-wide ice age of unimaginable proportions was now based on solid evidence. Kirschvink was converted. He coined the term "Snowball Earth" to describe the catastrophic deep freeze that had now, at least in his mind, become an historical fact. A disastrous glacial age had almost ended life on Earth 650 million years before our time. Simple, multicellular organisms that had been thriving for more than a billion years were decimated by this brutal 15-million-year glacial age. The world's oceans were capped with a layer of ice that was almost a mile thick at the poles and 65 feet thick at the equator. On an unusually warm day in the "tropics" of Snowball Earth, the temperature might have reached a torrid -30°C.

If, during this period, there had been a Martian civilization with astronomical telescopes and a space exploration program, they would never have bothered to send a probe to look for life on Earth. Our planet would have appeared as a small, brilliant white sphere through their telescopes — a barren, hostile desert that had been locked in ice for as long as our hypothetical Martian civilization had existed. No one could have guessed that life was hanging on precariously beneath the ice.

BACK TO THE DRAWING BOARD

But science is a skeptical, often conservative field. As Max Planck once grimly quipped, "Science advances one funeral at a time." Kirschvink found few allies for his resurrection of Harland's theory; his critics pointed out that the terrestrial albedo effect, which governs the reflectivity of the Earth and its ability to shed or acquire heat, had two corollaries that damned Harland's hypothesis. First of all, it was impossible for glaciers to reach the equator, and there was math to prove as much. Second, if glaciers had somehow managed to overrun the tropics and all the oceans had frozen, then terrestrial albedo dictated that there would be no way out of the deep freeze. It would last forever.

Kirschvink would not admit defeat. Yes, the balance of seasonal temperatures and terrestrial albedo seemed to rule out the possibility of glaciers extending any further south or north from the poles than 40° latitude (New York City or Tasmania). And certainly, at first, he couldn't coax any flexibility out of the numbers. But then he had a stroke of luck.

A Russian climatologist by the name of Mikhail Budyko had been studying the effects of nuclear winter. Years before, in the 1960s, Budyko had discovered a terrifying formula, a climatic tipping point that kicked in when half the sun's heat was reflected away from the Earth by the albedo effect of large glaciers. He proved mathematically that under certain climatic conditions glaciers could reach the latitude of 32° north or south (Houston, Texas, in the northern hemisphere or Santiago, Chile, in the southern hemisphere). This latitude was the point of no return; once a continental glacier crossed this line then runaway ice-albedo feedback would cause catastrophic heat loss. The result would be a planetary freeze. For Kirschvink, Budyko became the white knight of Harland's theory. Kirschvink contacted him, and within weeks Budyko's calculations were in his hands. The math was airtight.

Kirschvink was elated. The first flaw had been overcome. Unfortunately that left an even larger question: How the hell did the planet get out of an irreversible deep freeze? How did life survive? Kirschvink wrestled with the problem for months. There had to be an outside force, some vector of climate change he hadn't thought of. An asteroid impact like the one that wiped out the dinosaurs was a possibility, but dust and ash kicked up by the collision would have cooled the atmosphere even further. There was no evidence of greater solar activity either; if anything, the sun was slightly dimmer than it is today.

The answer had to be volcanoes — volcanoes with a series of eruptions of epic proportions. On average, volcanoes pump 10 billion tons of carbon dioxide into the atmosphere a year. If several supervolcanoes erupted at once, the amount of carbon dioxide would increase a thousandfold. Luckily for us, rain washes volcanic carbon dioxide out of the air. But since there wasn't any rain, since all the water on the planet was frozen, the carbon dioxide would not be "scrubbed" (as they refer to the carbon dioxide removal process on the space station). Kirschvink did some quick calculations on ratios of carbon dioxide and greenhouse warming and finally realized exactly how Snowball Earth came to an abrupt, and cataclysmic, end.

These days we don't need to be told how potent a greenhouse gas carbon dioxide is. Even at current levels, way less than 1 percent of our atmosphere, its warming effects are felt. Kirschvink's calculations revealed that after 10 million years, carbon dioxide levels had risen to occupy 10 percent of Snowball Earth's atmosphere. No matter how cold the planet, the greenhouse effect of carbon dioxide at such high levels was irresistible. And that was how the snowball melted. But there was more. The transition from snowball to hothouse was extraordinarily abrupt. The evidence suggested that the planet went from an average temperature of -40°C to an average temperature of 23°C in less than a few hundred years. Now all Kirschvink needed was proof of the great meltdown. It was 1990.

THE FINAL HURDLE

In 1992, Harvard geologist Paul Hoffman met Kirschvink and became an instant convert to the theory. Hoffman said he knew the very place to look for evidence: Namibia. Hoffman phoned an old friend, a geochemist by the name of Daniel Schrag, and convinced him to come along on an expedition to the 600-million-year-old calcium carbonate cliffs in Namibia.

Geologists had never known what to make of these geological

formations, but now, armed with the Snowball Earth meltdown hypothesis, Hoffman and Schrag had a good idea. They found the layer of dropstones precisely where it should be, at the 600-million-year-old level, and immediately above it, they found massive calcium carbonate deposits. Schrag's chemical analysis of these deposits left no doubt: they had been formed by calcium leached out of surface rock by carbonic acid. Here was proof that the vast amount of carbon dioxide in Snowball Earth's atmosphere had bonded with the first rain to fall for 15 million years and turned it into carbonic acid. The carbon in the acid rain then bonded with the calcium from the dissolved rocks to form calcium carbonate. There had indeed been a meltdown.

And what a meltdown. Hoffman was awestruck by the obvious speed of the transition; there was no gradation between the dropstone layer and the calcium carbonate layer. The heat wave must have come on so fast that it turned the Earth's climate upside down. Continental glaciers melted within decades. Some scientists believe that giant hurricanes churned the hot oceans, whipping up storm waves 330 feet high. The torrential rains unleashed by this "mother of all greenhouse effects," as Hoffman put it, probably lasted without interruption for a century. This scenario was the last apocalyptic piece of the puzzle.

In 1998, Hoffman and Schrag went on a world tour of major universities to expound their theory of Snowball Earth. It all went well until they hit another seemingly impenetrable brick wall. This time, the critics were biologists. Life could not have survived such a cataclysm, the biologists said. Photosynthesis required sunlight and how could sunlight penetrate ice that was more than 65 feet thick? Life only existed in the oceans at that time, so without open water, life wouldn't have survived. The Snowball Earth thesis was kicked to the curb again.

But there was one more white knight, a scientist who didn't just think out of the box but off the planet. Chris McKay was a planetary exobiologist working for the space science division of NASA. His corner of expertise was the survival of life in hostile environments, such as might be found on other planets. When he heard about the biological objections to the Snowball Earth theory, he already knew the answer. He had been to the one place on Earth with the same conditions: Antarctica.

NASA had been studying the dry valleys near McMurdo Sound in Antarctica because the terrain in these valleys is similar to Mars — a cold, snowless, rocky desert. Two of the valleys contain lakes that are millions of years old. And there, under ice more than 16 feet thick, is water. If life had survived beneath that ice, then it probably would have survived under the thicker ice of Snowball Earth.

Divers had been exploring the icy waters of both Antarctic lakes since the early 1970s and had discovered the same ancient form of cyanobacteria that predated Snowball Earth. Lots of it. As well, they found mats of algae as big as sheets. Another surprise was the amount of light penetrating the thick ice. McKay, an expert on polar ice, knew that when ice forms slowly, under extremely cold conditions, it is much more transparent than ordinary sea ice. So he didn't doubt that enough sunlight had penetrated the equatorial ice on Snowball Earth to sustain photosynthesis. Harland's theory had finally passed its last test. Snowball Earth had made it into the official geological record of our planet.

Life, hardly given the warmest of welcomes, survived. Bacteria, algae and anaerobic organisms living near geothermal deep-sea vents and in hot springs, continued to exist for millions of years until, in the great thaw, the icy clutch of winter was broken. The cosmic spring that followed witnessed life's resurrection in a renaissance that would change the face of the Earth.

Of course, when such a counterintuitive theory got the official nod, then any scientist with a hankering for posterity (and there are a lot) began looking for other, older glacial ages. And they found them. In the four-billion-year history of our planet, ice ages have struck at least six times, perhaps seven. Harland, Kirschvink and Hoffman's Snowball Earth now has the official name of the Marinoan glaciation. (I don't know why, after all their work, it didn't end up being called the Hoffmanian.) It was preceded by the Sturtian glaciation, which lasted much longer (720 to 660 million years ago), although it was not as global as the Marinoan or the even earlier Huronian glaciation, which gripped the planet 2.3 billion years ago, after the first oxygen hit the atmosphere.

The causes of some of these glaciations are still unknown. We know much more about the Pliocene-Quaternary glaciation — the one that we are in the middle of and the one that almost wiped us off the planet 200,000 years ago.

11

CLIMATE CHANGE PAST AND PRESENT

THE ARCTIC EOCENE

An ice age is not normal. During the 4.5 billion years of our planet's existence, only 517 million years have been spent in ice ages. That's less than a 10th. Earth's default climate is subtropical. Normally, palm trees grow as easily in the Arctic as they do at the equator. Up to 2.6 million years ago, when our current glacial age began, we had had a pretty long run of hot weather: 257 million years worth. Indeed, at the height of that extended summer, a time called the Eocene

Optimum, which peaked about 45 million years ago, the planet was particularly warm. Carbon dioxide and methane levels rose precipitously at the beginning of the Eocene Optimum and then stabilized at those levels for millions more years. Antarctica was fringed with subtropical rainforest, and palm trees grew in Alaska. Meanwhile, the Canadian high Arctic was home to vast forests of dawn redwoods and palmettos inhabited by an extraordinary menagerie of creatures. Fossils of tapirs, boa constrictors, rhinoceroses, pygmy hippopotamuses, giant land tortoises, eight-foot-long monitor lizards, alligators and the six-foot flightless "terror bird" called diatryma have been found on Ellesmere Island in the Canadian Arctic.

Due to their northerly latitude (polar landmasses were less than five degrees from their present positions), the Arctic forests experienced months of midnight sun and months of Arctic night. But even during the protracted darkness, the temperature never dipped below freezing. Some evolutionary biologists think that this might have been where some nocturnal mammals first evolved, prowling the warm, polar nights. Fossils of gliding lemurs have been discovered on Ellesmere, and perhaps the first bats found their wings under similar conditions.

So when the first snow fell, it must have come as rude awakening. Indeed, here is reason to believe the transition from the endless global summer to our current ice age was quite abrupt. The fossils from the forests of Ellesmere Island aren't even fossilized, despite being millions of years old. The mummified wood burns easily and the leaf litter from the floor of the ancient forest is still spongy. It is as if the whole forest had been freeze-dried *in situ*.

THE FIRST WINTER

During the Eocene, and through the following Miocene and Pliocene epochs, the continents continued to drift as the inexorable subsurface vortices of magma pushed the majority of tectonic plates northward. Antarctica was the wayward plate, drifting in the opposite direction, toward the South Pole. About 40 million years ago, five million years after the Eocene Optimum, Australia separated from Antarctica, and cool ocean currents set up between the two continents. For the first time snowflakes fell in the Antarctic highlands during the winter, although the rest of the continent still enjoyed a subtropical climate. Then, roughly 23 million years ago, during the Miocene epoch, the narrow isthmus of land separating South America from Antarctica was breached by the Drake Passage. Now Antarctica was completely isolated in an increasingly cooler southern ocean. The glaciers that had formed on the Antarctic mountain peaks began to spread, crushing the austral forests before them until, 15 million years ago, the glaciers were well on their way to covering the entire continent.

The rest of the world still enjoyed the glow of the Miocene epoch, almost as warm as the Eocene, but it was like being in a house where somebody has left an air conditioner on full blast, day and night, in a spare room. It was just a matter of time before something tipped the global climate into an ice age. And sure enough, that triggering set of conditions occurred 2.6 million years ago, causing the climate to cool too quickly for the subtropical forests of the Arctic Pliocene epoch to adapt. Some evidence suggests a catastrophic, fairly rapid transition from subtropical to ice-age temperatures as the first of 11 glacial periods, now known as the Pre-Illinoian glaciations, established ice sheets in the northern hemisphere and initiated the Pliocene-Quaternary glaciation. Global temperatures plummeted and ocean levels dropped as water was locked into vast ice sheets.

Hominids had just gotten a foothold in Africa at the beginning of the Pliocene-Quaternary glaciation, with two separate hominid lineages: *Australopithecus africanus* and *Homo habilis*. It was probably the latter that eventually led to *Homo sapiens*. Fortunately for us, Africa was spared the more brutal climatic effects of the

Pliocene-Quaternary, and the deterioration of world climate held off for the critical gestation period of our species, at least until the onset of the Illinoian glaciation, 191,000 years ago.

MESSAGE IN A BOTTLENECK

With almost seven billion humans currently alive on our planet, it might seem hard to believe that we were once on the verge of extinction. Yet just after the first anatomically modern humans, *Homo sapiens*, arose in Africa, the climate took a big turn for the worse. The warm interglacial temperatures that had nurtured our forebears grew much cooler, and food became scarce. The cradle of our species was tipping us out. This cold snap, the full-blown glacial advance of the Illinoian period (referred to by climatologists as Marine Isotope Stage 6 after the ocean bottom sediments laid down at that time) began about 195,000 years ago and lasted, with a few brief interglacial periods, for about 72,000 years.

According to paleoanthropologist Curtis W. Marean, a professor at Arizona State University, central Africa became virtually uninhabitable, and the only safe haven for our ancient ancestors was the sea coast of South Africa. Ocean levels had dropped more than 330 feet, but here, on the coast, plentiful marine life and edible shore plants tempered the hard, cold millennia. Even then, it was tough going. During a particularly severe period, as the glaciers advanced to their maximum extent, Marean postulates that our species dropped from more than 10,000 individuals to a just few hundred souls. This population choke point left a telltale genetic imprint in our genes. Geneticists discovered it in the early 1990s, like a message in a bottle that had been afloat on the cellular seas of time, a tale from our past selves to our present selves, describing in DNA code a harrowing account of near extinction.

The only other advanced hominids at the time were Neanderthals. And the paleoanthropological consensus seems to be that the Neanderthals likely wouldn't have survived. Certainly there is archeological evidence that Neanderthals had fire, clothing and culture, but their physiology wasn't as efficient as *Homo sapiens*. If Neanderthals had failed, then the great hominid evolutionary experiment might have failed as well. We would have a world today without cities, without civilization, without the "we." It would be a planet where mammoths still browsed on young birch trees in Minnesota and Norway, where marsupial wolves still prowled the Australian outback and where dodos strutted through the undergrowth on the island of Mauritius without fear of dying out.

The Illinoian glaciation ended with an interglacial period now known as the Sangamonian stage, about 130,000 years ago. And then it was off to the races for *Homo sapiens*. Earlier, during the hard winter of the Illinoian glaciation, *Homo sapiens* had acquired the crucial traits of prosocial cooperation and projectile weapons, essential skills for our survival through the dark times. So we were ready when the climate warmed up and megafauna returned to the plains of Central Africa. The Sangamonian summer and autumn lasted some 50,000 years, from about 125,000 to 75,000 years ago, and we used it well. Our species expanded rapidly northward, colonizing the whole of the African continent. Around 70,000 years ago, we spilled out of North Africa, just as one of the bitterest glacial ages ever, the Wisconsin glaciation, was beginning. But that didn't slow down our global expansion.

By the time the Wisconsin glaciation entered its most severe period, or last glacial maximum (LGM) about 26,000 years ago, our restless, nomadic species had spread to the far corners of the planet — from northern Europe and throughout Asia to Australia, as well as North and South America. Our colonization of the world took place during the worst ice age since the Andean-Saharan ice age, 460 million years before that. Over the cold, dark millenia of the Wisconsin ice age, we developed complex languages and culture and religion. The Lascaux cave paintings were executed 17,300 years ago at the height of the LGM. Rendered by flickering torch-light on limestone walls, these exquisite paintings of ice-age mammals are snapshots of a lost era when Europe was either covered by Arctic tundra or buried under glaciers. Two-mile-high cliffs of continental glaciers were parked just north of present-day London, England, a mere 435 miles north of Lascaux, and world sea levels were 330 feet lower than today.

Yet less than 7,000 years later, the Wisconsin glaciation came to a surprisingly sudden end, and our globetrotting forebears had to endure a series of catastrophic climate changes. Greenland ice cores reveal that the end of the Wisconsin was a period marked by intense climatic instability. The mean global temperature flipped several times between temperate conditions and those of an ice age. When the global melt was really underway, about 15,000 years ago, Greenland's average temperature shot up by 16°C in a period of 50 years, perhaps fewer. And around 12,000 years ago, the definitive end of the Wisconsin, Greenland's mean temperature skyrocketed by 15°C in one decade. As a result, the reset global temperature, the new "normal," was probably 6°C warmer. Climatologists now refer to this as "abrupt climate change." Imagine if this happened today. None of our low-lying nations would be able to build levees and sea walls fast enough to stave off an ocean rising more than three feet a year.

By the time the oceans peaked, when the Wisconsin glaciers finally retreated to their present positions some 11,000 years ago, humans were thriving on every major continent except Antarctica. The cultural and technological advances we had been nurturing through the long, dark millennia of the Wisconsin glaciation blossomed into grand civilizations. It was the global spring of our present era: the Holocene interglacial.

In that 11,000-year period, from the retreat of the continental glaciers to the present, humans have advanced quickly: the first agriculturally based city states arose 6,500 years ago, the domestication of the horse along with the first written language 6,000 years ago, the invention of the wheel 5,400 years ago. Iron was first smelted about 3,200 years ago, and 2,400 years ago Euclid discovered geometry, the basis for modern mathematics. The pace of invention accelerated as technology was increasingly harnessed to advances in science, and today humans stand on the brink of an extraordinary destiny. Maybe Carl Sagan was right when he said, "We are a way for the cosmos to know itself." Certainly we are coming to understand climate and, in particular, we are making great progress in decoding the mysterious engine that drives climate change. And thanks to Milutin Milankovitch, we are even beginning to understand the cycle that drives the ice ages themselves.

THE MILANKOVITCH CYCLE

Milutin Milankovitch (1879–1958) grew up in the village of Dalj on the banks of the Danube on the eastern edge of the Austro-Hungarian empire. The eldest of seven siblings, Milankovitch was only eight when his father died. His mother, stranded with six children under the age of seven, enlisted Milankovitch's grandmother and uncle to help with the young family and continued the home schooling Milankovitch's father had begun. Relatives and friends, some of whom were prominent inventors, philosophers and poets, tutored Milankovitch as well. It was a turbo-charged education, and he was more than prepared when, at age 17, he enrolled in the Vienna Institute of Technology.

Milankovitch became one of the many geniuses whose lives unknowingly paralleled each other's during Vienna's astonishing fin de siècle. He graduated at the top of his civil engineering class and, after being awarded his doctorate, began work at a Viennese engineering firm. In his spare time, he registered a series of successful patents that quickly made him wealthy, so much so that by 1912, at the age of 33, he was able to indulge his personal interests on a number of fronts. One of these was the orgin of ice ages.

In June 1914, just as war broke out, he married Kristina Topuzovich. They decided to spend their honeymoon in his home village of Dalj. As it happened, Dalj was in territory contested by Serbian nationalists, and Milankovitch, having been born there, was arrested by the Austro-Hungarian army and thrown in jail. He wrote in his diary, "The heavy iron door closed behind me . . . I sat on my bed, looked around the room and started to take in my new social circumstances." Fortunately, the soldiers allowed him to keep his briefcase, in which he had hurriedly stuck his theoretical papers and some blank sheets. "I looked over my works, took my faithful ink pen and started to write and calculate . . . When after midnight I looked around the room, I needed some time to realize where I was. The small room seemed to me like an accomodation for one night during my voyage in the Universe."

Milankovitch's wife appealed to their highly placed social connections in Vienna to get him transferred to a military prison in Budapest, where his captivity was more lenient, a sort of house arrest with day privileges. He spent the balance of the war researching his climate theories at the Central Meteorological Institute in Budapest, and it was there that he began to investigate the extraterrestrial causes of ice ages. When the war ended, he rejoined his family in Belgrade and started work on his crowning theory: a mathematical description of Earth's climatological history, published in 1924 as *Climates of the Geological Past.* His thesis is now known as the Milankovitch cycle, which, at its complex heart, is an interplay of three variables that coincide every 100,000 years to create an ice age.

The first variable is the Earth's axial tilt, which is not fixed. It gradually alternates 2.4° (between 22.1° off the vertical and 24.5° off the vertical) over a period of 41,000 years. What produces the tilt are tidal forces exerted by the moon and nearby planets. Right now, the Earth's tilt is 23.44°. The Tropics of Capricorn and Cancer are respectively 23.4° below and above the equator, while the Arctic and Antarctic circles lie the same distance away from the poles. If Earth was tilted any more obliquely, say at its maximum of 24.5°, then the Tropic of Cancer would also be shifted north significantly, to 24.5° latitude and instead of running through Havana, it would run through Miami. Likewise in the Middle East: instead of running through the United Arab Emirates, the Tropic of Cancer would run through southern Iran. The same expansion would apply to the southern hemisphere, so the overall area of the tropics would increase by more than 2° latitude. This expansion would come at the expense of the temperate zone, because the Arctic and Antarctic circles would also creep south and north, enlarging the polar regions. Any change in the Earth's tilt affects the severity of the seasons. When the Earth is more vertical, summers get cooler in the northern hemisphere and this, more than cool summers in the southern hemisphere, favors the onset of an ice age. Why?

It's a matter of landmass and water. Most of the Arctic is covered by ocean, which means it can't get as cold as the Antarctic: the Arctic Ocean, despite its icy cover, acts as a reservoir of heat in the form of liquid water. The Antarctic, being solid land, has no such thermal check on how cold it can get. As a result, the northern hemisphere keeps our planet warm by counterbalancing and compensating for the southern hemisphere, which is cooler, on average, than the northern hemisphere. Just ask the someone living in the Falkland Islands. Although they're on the same latitude in the southern hemisphere as London, England, is in the northern hemisphere, the Falklands have an average yearly temperature of 5.6°C whilst in London, the average is twice as warm, 10.4°C. Again, water drives the thermal imbalance here but in the opposite direction. Oceans may take longer to cool down, but they also take longer to warm up. The larger ocean surface of the southern hemisphere drives the overall temperature lower, even in the summer. Consequently, Earth has one foot in the South Pacific ice bucket, and the only thing keeping the planet warm is summer in the northern hemisphere.

So let's fastforward Earth's tilt 20,000 years to its most upright inclination, 22.1°. The summers will have been cooler in the northern hemisphere for thousands of years. Yet the glaciers haven't started moving yet. Why? Well, there are two more things that need to happen at the same time to force Earth's climate into an ice age. One of these, the second variable in Milankovitch's trio, is the seasonal timing of our orbital perihelion, the point at which Earth's orbit brings it closest to the sun. Right now that happens in January, when the southern hemisphere is tipped toward the sun. But over thousands of years, the point of perihelion moves. This means that in 13,000 years, the northern hemisphere will get the summer perihelion boost. This also means that one of the three requirements necessary to tip our climate into an ice age is currently in place. The northern hemisphere has lost the extra sunshine that the perihelion would give it, and the southern hemisphere simply squanders the extra heat — a 7 percent increase in solar energy — with its summer chilliness. Fortunately, the northern hemisphere can get along without the extra encouragement from the perihelion.) But when the perihelion cycle and Earth's tilt coincide with the third and most important cycle — orbital eccentricity — then it's time to get out the snow shovels and road salt.

Every 96,000 years or so, the Earth's orbit switches from a nearly circular orbit to a more elliptical one. This changes everything. During a more elliptical orbit, the perihelion brings the Earth much closer to the sun and increases the solar energy received by a whopping 20 to 30 percent on the hemisphere facing the sun. You'd think that would warm things up considerably, but the good news is counterbalanced by the farthest point that an elliptical orbit would take Earth away from the sun, the aphelion. If the aphelion occurred while the northern hemisphere was tipped away from the sun (as it is now), it would cancel out the increase in solar radiation on the southern hemisphere. Every winter would add to the snow cover that remained over the summer in the northern hemisphere, because landmasses lose heat more quickly, particularily if the Arctic sea ice remained year round. Combine that with a less oblique axial tilt and, presto, you have an ice age underway.

Fortunately for us, that that won't happen for another 60,000 years. Although there are still some scientific critics of Milankovitch's cycle, the geological evidence from the Vostok ice cores and deep-sea deposition cores show that past ice ages are in lockstep with his axial precession index and orbital eccentricity charts. There seems to be little question that, at least over the past 2.6 million years, during the Pliocene-Quaternary glaciation, Milankovitch's cycle has been driving Earth's climate on the macro scale.

But even within these large, regular cycles, there are climatological events that are highly irregular and reveal the sensitivity and potential instability of the atmosphere. The advance and retreat of the glaciers and the beginning and ending of warm interglacial periods are not always orderly and gradual. A stressed atmosphere can be nudged out of balance by extraordinarily subtle influences.

CLIMATE INSTABILITY

When climates are stable, whether in the deep freeze of a Snowball Earth or the tropical planetary conditions that prevailed for millions of years between ice ages, they have climatic equilibrium. When instability is introduced, some outside influence, say by the Milankovitch cycle or from a change in the percentages of atmospheric gases, then the climate goes into a period of wild instability. Unusual or extreme weather events sometimes accompany the state of transition between the prior set of conditions and the next. As we've seen, the evidence from the ice cores in Greenland suggest tipping points that usher in apocalyptic climate change.

Sometimes the early warning signs of change are registered by differences in the behavior of animals and plants. Like canaries in a coal mine or dogs barking before an earthquake hits.

PHENOLOGY . . .

I'm a naturalist. It might seem a contradiction, living as I do in urban Toronto, but you'd be surprised at the wildlife that calls downtown home. I keep a diary of the most interesting urban creatures that I see, as well as an annual "first sightings" chart, in which I record the day the maple leaves open — a robust indicator of the first warm weather — and when I first see my four favorite seasonal creatures: June bugs, nighthawks, swallowtail butterflies and cicadas. The study of these seasonal appearances (and disappearances in the fall) is called phenology and is, I suspect, *the* cardinal trait of the amateur naturalist. I'm in good company. I recently discovered that Thomas Jefferson was an obsessive phenologist. He kept a record of the difference in leaf-opening times between his estate in Virginia and his home in Washington. Over the course of the last 30 years, my charts have not only become a calendar of migratory and seasonal animals, they have also turned into a personal record of climate change. Contrary to what you'd expect, it seems that spring is arriving just a little later each year, at least it is in the North American Great Lakes region.

The most telling data comes from my leaf-opening records. During the 1980s, the maple leaves opened on April 23, but in the 1990s that shifted to April 29, a forward jump of six days. In the first decade of the new millennium, the average leaf-opening date remained at April 29, but in the first eight years of the second decade, from 2011 to 2018, the date moved a little farther, to May 1. Interestingly enough, that recent average includes a wildly early year, 2012, when the leaves opened on April 14.

So what's happening? Every decade, spring is getting colder and winter is lasting just a little longer, at least in Toronto. Perhaps it's just the maples. Certainly rising levels of carbon dioxide are influencing the atmosphere, but as it turns out there are other, human-originated factors that have equally potent effects on climate.

. . . AND CONTRAILS

Andrew Carleton, a geography professor at Pennsylvania State University, and David Travis, a climatologist at the University of Wisconsin, had always been curious about the influence of high-altitude air traffic on climate. What effect did jets have on the weather? Short of clearing the sky of all aircraft for several days — something they could never arrange — how would they be able to quantify any effect?

Then came 9/11 and the attack on Manhattan. For the three days that the U.S. grounded all commercial aircraft and almost all military flights, Carleton and Travis scrambled to amass the data that poured in from national sensors. When the numbers were gathered and crunched, the figure was surprising. After factoring out local weather and other thermal irregularities, they discovered that during the three-day, no-fly period after 9/11, the mean surface temperature of the United States climbed 1.2°C. Why? The simple answer: less cloud cover.

When jets fly at high altitude, their exhaust creates long streamers of ice crystals called contrails. Everyone is probably familiar with these twin white lines inscribed in the blue, and if the dew point is just right, they last well after the plane has disappeared over the horizon. Given how narrow contrails are, you'd think they'd be negligable in terms of cloud cover, but apparently, if there are enough of them, they contribute significantly to the continent's shade. Carleton and Travis published their results in *Nature*, and a new, somewhat contradictory term, "global cooling," was introduced into the climate-change lexicon. Perhaps that explains why spring seems to be arriving a little later every decade, at least here in eastern North America. But there are other ways that humans have been influencing the climate, and for much longer than you'd think.

ANTHROPOGENIC WARMING

Presently we are in the Holocene epoch, which began 11,500 years ago at the beginning of our current interglacial period. By some estimates, the last interglacial period, the Sangamonian, lasted 11,000 years. Does that mean our own interglacial period is past its expiration date? Some climatologists think so — University of Virginia emeritus professor William Ruddiman for one. He believes that our current interglacial period should have ended thousands of years ago and we should be in the throes of a new glacial advance. But we're not. Why? Ruddiman blames humans.

More than a decade ago, when Ruddiman was studying ice-core data from both Greenland and Antarctica representing thousands of years of atmospheric history, he noticed a spike in carbon dioxide levels that began about 8,000 years ago. Cross-referencing it with archeological records, he noticed a correlation. The carbon dioxide rise came at exactly the same time as slash-and-burn agriculture spread from the Middle East to Europe and western Asia. Then, about 5,000 years ago, there was another spike, this time in atmospheric methane levels. These Ruddiman linked to rice paddy agriculture, which was beginning in the Lower Yangtze region of China. The methane spike continued to rise as rice paddies expanded across the rest of China and into Asia. By 3,000 years ago, the combined effects of anthropogenic methane and carbon dioxide had elevated mean global temperature by 0.8°C at mid-latitudes and by 2°C at far northern latitudes. According to Ruddiman, this increase was large enough to have stopped a glaciation of northeastern Canada that, according to his calculations, should have started 2,000 or 3,000 years ago. "Whew!" you might say. "We dodged that bullet." But things climatic aren't straightforward. It's quite possible that the Ruddiman effect is now shifting into high gear.

Since the onset of the Industrial Revolution in 1800, human contribution to atmospheric carbon dioxide has been increasing almost exponentially, primarily from burning fossil fuels. This process has released billions of tons of carbon dioxide (that had been previously sequestered over hundreds of millions of years by the carbon cycle) into the atmosphere.

Today carbon dioxide represents only 0.04 percent of our atmosphere, or approximately 400 parts per million. This is a tribute to the sequestering efficiency of the carbon cycle, considering that ocean plants, mostly plankton and algae, along with land vegetation, pump 771 gigatonnes of carbon dioxide into the atmosphere yearly. By comparison, our human contribution of 29 gigatonnes annually seems trivial, but the carbon cycle is finely tuned, and there might be little capacity for excess. In fact, most climatologists believe that the carbon cycle has already been overwhelmed, and they argue that is reason why measurements of atmospheric carbon dioxide are currently rising.

Clearly the world is no longer evolving as it would have without our intervention. Here, in the midst of a catastrophic man-made mass extinction where thousands of species are being lost due to human activities, we also appear to be altering the very atmosphere itself. In light of this, a number of scientists contend that we are entering a new geological era of our own making: the Anthropocene.

TICKLING THE DRAGON'S TAIL

When Robert Oppenheimer and General Leslie Groves were overseeing the construction of the first atomic bomb in Los Alamos, New Mexico, they set up an experiment to measure the critical mass threshold of the bomb: the point at which a runaway fission reaction would create an explosion. The experiment was rather primitive. A horseshoe-shaped stack of plutonium bricks, just slightly below critical mass, was arranged on a tabletop amid an array of radiation detectors. A little rail entered the opening of the horseshoe and stopped at the center of the pile. This rail was the track for a sliding rod, and at its tip the scientists placed a small chunk of plutonium, just enough that, when added to the pile, it would trigger a critical mass reaction. The idea was to slip the chunk in and out so quickly the pile wouldn't get a chance to go completely critical and explode. That way they could measure the spike in radioactivity and fine-tune the mass necessary to detonate the first nuclear bomb. Rather aptly, they called this manoeuver "tickling the dragon's tail."

It seems that anthropogenic atmospheric change is a similar experiment, though unsupervised, a kind of meddling with the dials on our global thermostat that might have disasterous and unpredictable consequences. Our atmosphere is a fluid system, prone to turbulence and the butterfly effect. In July 2012, for example, a warm air mass parked over Greenland for 12 days, and on one of those days, 11 inches of ice melted, representing 14 percent of the total annual melt. It was a weather anomoly, an unpredictable event. Another small imbalance, like the Greenland heatwave, seemingly trivial, might cascade up the chain of complexity and eventually destabilize the entire global system. It has happened before.

CLIMATE AND WEATHER

"Climate is what you expect, weather is what you get."

ROBERT A. HEINLEIN

People often attribute extreme weather events to climate change, and although it's true that one of the harbingers of abrupt climate change can be extreme weather, it's best not to lose sight of the fact that it's the averages, the mean temperatures and rainfalls and wind speeds that indicate real climate change.

But when, or where, does climate stop being climate and become weather? We can say that certain types of weather are endemic to particular climates. A hurricane is a major weathermaker but is indigenous to the tropics. Occasionally, the remnants of a hurricane will travel beyond the 40th parallel but never all the way to the Arctic Circle. Conversely, blizzards never strike the Amazon.

The fact is weather never exceeds the bounds of climate; it is a subset of climate, a particularity. You could say that weather is the quotidian particulars of climate. The difference between the two is primarily time based. Weather has short duration, hours usually, days occasionally, sometimes weeks, and even, when a big volcano erupts, years. Climate takes decades and more often centuries.

Climate change has had extraordinary effects on human civilization over the last 10,000 years. It induced a famine that racked Egypt 4,200 years ago. It caused the desertification of the American southwest 800 years ago, which forced the Ancestral Puebloans from their elegant cliff cities. At the beginning of the Little Ice Age, 500 years ago, fickle monsoon rains collapsed the great civilization of Angkor Wat in Cambodia while, on the opposite side of the planet, the Viking settlements were frozen out of Greenland.

There was a cultural upside to the Little Ice Age though. The legendary sound of Stradivarius's violins owe much of their tonality to the 400-year-long cold snap. Why? Because Stradivarius used Croation maple wood for the neck and back of his instruments, and the particularly severe winters of the Little Ice Age slowed the maple's growth and compacted its wood. The long, icy winters in the mountains of Croatia imparted a unique, ethereal tone to the Stradivarius violin. Certainly climate change and decades-long droughts have changed human destiny. But history has also been changed by just plain weather, and sometimes much more profoundly than climate.

12
WEATHER THAT CHANGED HISTORY

"Fortune, which has a great deal of power in other matters but especially in war, can bring about changes in a situation through very slight forces."

JULIUS CAESAR

Never does the fate of a nation hang in the balance so delicately as during a pitched battle for its very survival, and, on a surprising number of occasions, weather has tipped the scales in favor of one side or the other. Weather was certainly the deciding factor in the fifth century BCE, when the Persians invaded Greece.

King Darius (550–486 BCE), the third king of the Persian Achaemenid empire, ruled at the peak of its power. His realm stretched from present-day Turkey in the west to the border of India in the east and from the Black and Ural seas in the north through the whole of the Middle East to Egypt in the south. It was on his watch that the Achaemenid empire began to annex Greek colonies, most notably in Ionia. But the Greeks did not take well to occupation.

A series of rebellions ensued, inflicting so much damage on Persian forces that in 490 BCE King Darius ordered the invasion of Greece proper. He assembled an army of 10,000 immortals (elite infantry), 10,000 light infantry, 5,000 archers, 3,000 cavalry and 600 naval vessels (triremes) and placed them under the command of his best generals, Datis and Artaphernes.

The Persians razed the island of Naxos and then landed on the Greek mainland where they besieged and destroyed Eretria. They then headed to Marathon. But here the Greeks took a stand. Miltiades, the Greek general, had only 9,000 Athenians and 1,000 Plataeans against the entire Persian army, and yet by the end of the day the Persians were routed, losing more than 5,000 soldiers. By contrast, Athenian and Plataean casualties numbered less than 2,000.

The victory at Marathon was a huge morale boost for the Athenians, especially since they had achieved it without Spartan help. For his part, King Darius swore to revenge this humiliating defeat, and the Persians withdrew in order to draw up new invasion plans. But King Darius died four years after the battle of Marathon and it fell to his son Xerxes to uphold his father's pledge. Xerxes invaded Greece in 480 BCE with a huge force. According to modern estimates, his army alone totaled somewhere in the neighborhood of 100,000 to 150,000 soldiers against an accumulated Greek standing army of about 52,000 souls.

Themistocles, the Athenian politician and naval strategist, was the general in charge of the Greek defense. His battle plan was to bottleneck the Persian invasion on land at Thermopylae and on water at the Straits of Artemisium. Both battles were stopgap measures, meant to hold off the Persians rather than defeat them. Here is where the weather comes in.

Several days before the great naval engagement in Artemisium, which historians believe took place sometime in August or September 480 BCE, the Persian fleet was caught in a terrific gale off the coast of Magnesia near Thessaly. Four hundred of their 1,200 triremes sank, a staggering loss. (Persian triremes were not inconsequential vessels. Like the Greek triremes, they had been perfected over hundreds of years of naval warfare and each 40-ton trireme was more than 120 feet long with 120 oarsmen.) Yet, with 800 triremes left, the Persians still had a massive logistical advantage. Their commander ordered 200 vessels to sail around the coast of Euboea in a maneuver to entrap the Greek navy, but against all odds another storm struck, and all 200 were shipwrecked. The Persians had lost half their fleet to bad weather.

Even this monumental loss wasn't enough to save the Greek fleet. A few days later the two navies met at Artemisium, and the odds were still in favor of the Persians. The 270 ships that the Greeks mustered were vastly outnumbered, and in the ensuing fray both navies lost a hundred vessels. The Greeks retreated with the remainder of their fleet.

Meanwhile, as the Greek and Persian ships clashed on the seas, a historic battle was taking place on land at Thermopylae. This time the Spartans had decided to join the fray. A Greek force of about 7,000 soldiers led by the Spartan king Leonidas held off the entire Persian army on Thermopylae's narrow strip of coast. The battle would have been a stalemate had not a local resident named Ephialtes betrayed Leonidas by disclosing the location of a secret path that led behind Greek lines. When Leonidas got wind of the treason, he realized his position was hopeless. He dismissed most of the army and remained with a diversionary force of 300 Spartans, 700 Thespians and 400 Thebans. This final, desperate battle became legendary. By the time the last Spartan fell, the Persians had lost 20,000 men.

After their costly victories in Greece, much of the Persian army withdrew to Asia. Xerxes left behind his trusted general, Mardonius, to complete the conquest with a smaller force. Mardonius's army went on to sack Athens (which had been evacuated), while on the Aegean Sea the Persian navy chased the Greek ships until, late in the autumn of 480 BCE, they cornered the Greeks in the straits at Salamis. But this was a trap that would backfire for the Persians. Themistocles knew these waters well, and he knew the autumn weather even better. He had a meteorological trick up his sleeve.

Having no inkling of Themistocles's hidden ace, the Persian commander presumed that, with his 500 triremes to the Greeks' 370, he had effectively trapped and outnumbered them. The Greek vessels were on the other side of a headland from the Persians, not in direct line of sight. At sunrise on the day of battle, there were so many Persian boats they formed a solid line across the straits, blocking any chance of escape for the Greeks. The Persians waited, knowing the Greeks would never offer terms. Yet the morning wore on. Uncharacteristically, or so it must have seemed from the vantage of the Persians, Themistocles delayed action until mid-morning, when he sent out a handful of his ships. Instead of engaging the enemy, however, the Greeks turned and retreated to their redoubt. The Persians took the bait, and the entire fleet rowed into the narrow waters of the strait. They had no idea that the wind god Aeolus was about to enter the fray on the side of the Greeks.

As soon as the Persians rounded the headland, the Greeks launched their assault, ramming the Persian triremes with deadly accuracy. Several Persian vessels began sinking and chaos ensued — too many boats crammed in too small a space. Right on schedule, the late morning Etesian wind arrived, blowing the top-heavy Persian ships against each other and creating a tangle of boats.

The Greeks were able to outmaneuver and sink more than half the Persian fleet before it retreated in complete disarray.

It was the beginning of the end for the Persian campaign. The Persian army remained in Greece for another year before it was defeated at the battle of Plataea in August 479 BCE. Almost simultaneously, the vestiges of the Persian navy were crushed at the naval battle of Mycale.

The battle of Salamis turned out to be the turning point of the war, and the Persian navy never regained its former strength. If not for the dependable Etesian wind, Greece's great classical age would have been stillborn, and the course of Western civilization irrevocably altered.

THE FALL OF DACIA

"Extraordinary rains pretty generally fall after great battles; whether it be that some divine power thus washes and cleanses the polluted earth with showers from above, or that moist and heavy evaporations, steaming forth from the blood and corruption, thicken the air."

PLUTARCH

Domitian, who ruled Rome from 81 to 96 CE, oversaw the empire at the height of its power and territory. A ruthless, despotic tyrant, he was nonetheless popular with the legions after he led several successful campaigns in Britain. Yet he was never able to tame the Dacians, who maintained a huge territory (approximately the area now occupied by Romania and Bulgaria) less than 700 miles from Rome. They were constantly harassing Roman settlements. In 85 CE, they invaded the Roman province of Moesia and killed the governor, Oppius Sabinus. The Romans beat them back, but in 86 CE they invaded again. This time, Domitian sent a legion commanded by the Praetorian prefect Cornelius Fuscus. The two forces met at Tapae in a battle that ended with a humiliating defeat for the Romans. An entire legion was wiped out. Prefect Fuscus was killed, and the Dacians captured the Roman battle standard, or *aquila*.

While Rome could countenance a military defeat, which was sometimes unavoidable, the loss of their standard was too much. Domitian petitioned the senate to renew the campaign against the Dacians, but, like Darius before him, he didn't live to see Roman revenge. On September 18, 96 CE, he was assassinated, and it fell to the next emperor, Trajan, to get the job done.

Dacia became Trajan's personal vendetta and he invaded it in 101 at the command of two legions, penetrating all the way to Tapae and the same battlefield where the Romans had been defeated 15 years earlier. There he met his well-armed and formidable opponents.

The battle went well for the Dacians at first, but then a thunderstorm swept over the combatants. The Romans saw it as a sign that Jupiter had joined the battle on their side, and they fought in the rain with renewed vigor. Unwilling to take on both Jupiter and Trajan's legions, the Dacians retreated. The second battle of Tapae was a great success for the Romans, who then pressed their advantage north until the Dacian king, Decebalus, sued for peace. A summer storm had tipped the balance.

HAMMER OF THE NORTH

Bad weather can last for years. Such was the case in Norway, Sweden and Denmark during the sixth century when two massive volcanic eruptions caused a volcanic winter. The first is thought to be the Tierra Blanca Joven eruption in central El Salvador. Ash was carried into the upper atmosphere and then encircled the globe. In 536 CE, the year after the eruption, the Byzantine historian Procopius wrote, "During this year a most dread portent took place. For the sun gave forth its light without brightness, like the moon during this whole year, and it seemed exceedingly like the Sun in eclipse." The second eruption happened around 539–540, probably somewhere in the tropics. The bitter decade after these eruptions was the most extreme short-term cooling event of the northern hemisphere in the last 2,000 years. There were crop failures and famines worldwide.

Scandinavia was hit the hardest. Most of the farming villages in the Uppland region of Sweden were abandoned as the population succumbed to starvation. Similar scenes played out across the rest of Scandinavia. Those that survived did so by force, seizing the few remaining provisions at sword point. It was from this dark crucible that a new Scandinavian order arose: marauding bands of warriors, armed to the teeth, who had honed their martial skills by fighting each other for supremacy over the long, dark years. And when the summer finally returned, they took their new talent for plunder and mayhem on the road. Or, should I say, on the water. Viking longships sailed westward all the way to Newfoundland, south into the Mediterranean sea and as far east as the Volga River in Russia.

A RUSSIAN WINTER

In the summer of 1812, Napoleon invaded Russia with some 680,000 soldiers. By September, he had taken Moscow. Eerily, the capital was deserted; there wasn't a soul left to engage the invasion force. The Russians had evacuated the city before Napoleon entered, and they refused to sue for peace. Napoleon stayed on a month while his soldiers pillaged the burned-out ruins. In October, with a negotiated settlement out of his grasp, he began to chase the elusive Russian army southwest. That's when the weather began to turn. It must have dawned on Napoleon, too late, that the Russians had outmaneuvered him with their stalling and evasion tactics. In November, an early and exceedingly bitter Russian winter descended in all its frosty fury.

Temperatures dropped below -40°C. Expecting a short campaign, Napoleon's troops had been issued no winter clothing. Furthermore Napoleon's supply lines had been decimated by sporadic Cossack attacks, and the Russians had burned their crops to deprive the invaders of local food supplies. In one pitiless 24-hour period, 500,000 horses died from the cold. By the time his beleaguered army trudged out of Russia, Napoleon had lost 380,000 men to hypothermia and the Cossacks. One hundred thousand more had been captured. Napoleon's failure became France's failure, and it lost its primacy among its allies. Austria and Prussia switched sides, and the balance of power in Europe shifted dramatically.

Who knows how European history would have played out if Napoleon's army had not been destroyed? Had he but launched his campaign against Russia a few months earlier, the outcome would have been completely different.

THE FLANDERS OFFENSIVE

By 1917, the "war to end all wars," a conflict most military strategists thought would be over before Christmas 1914, had dragged on for three bloody years in a stalemate that cost hundreds of thousands of lives. In the spring of that year, British commander Field Marshal Earl Haig and his French counterpart, Robert Nivelle, planned a summer offensive on the German positions in Flanders. It would be a massive assault using all the technology available at the time: artillery barrages, tanks and thousands of troops. The plan was to overrun the German positions so quickly that, according to Haig, the allies would retake Belgium in three hours. They chose August to launch the campaign because it was invariably a month of dry, sunny weather in northern France. But the weather gods were not on the side of the Allies.

On July 21, rain began to fall on the front lines. It rained again the next day, and the day after that and the day after that. It rained continuously, for the whole of August. The offensive, which began on July 31, was quickly bogged down in the mud. Even the tanks were mired, and any dreams of taking Belgium, or even the village of Passchendaele, just inside the German lines, were lost to the wettest summer Europe had seen for 30 years.

The battlefield became a shell-blasted sea of muck where men literally drowned. There was a brief respite from the rain in September, but in October the rains began again, and Passchendaele wasn't taken until November, 15,000 Allied casualties later. The victory, at this point, did little to change the balance of power along the front. If the August deluge hadn't flooded the battlefield, the First World War might have ended much sooner.

THE MIRACLE OF DUNKIRK

By the Second World War the science of meteorology was indispensable to military strategy. The famous 24-hour postponement of D-day demonstrates the power of a single forecast. Yet the vagaries of weather had other surprises for the combatants. The improvised evacuation of the British troops at Dunkirk was certainly the first of these, although in this instance it was fog and lack of wind that saved the day.

In the spring of 1940, Hitler's lightning-fast war overran France so quickly that more than 400,000 Allied soldiers (Belgian, Canadian, British and French) were completely encircled by German forces on the north coast of France by May 21. Their only hope was a retreat by water. Admiral Bertram Ramsay and the leader of the British Expeditionary Force, General Gort, began to prepare a marine evacuation for the desperate troops.

There was little time, and the situation was looking grim when, mysteriously, they were given a precious two-day reprieve. On May 24, the German High Command issued a halt order; Gerd von Rundstedt, commander for the German army in France, was concerned that his panzers and forward troops might be vulnerable to attack. Besides, Hermann Göring had argued that air power alone could destroy the retreating Allied forces. This delay was critical for the Allied evacuation plan. The weather gods were also in their corner: at the same time as the halt order, the weather took a turn for the worse. The British admiralty swung into action. Under heavy cloud cover, they sailed toward Dunkirk undetected by the Luftwaffe.

On the afternoon of May 26, Hitler rescinded the halt order, and the German army began to advance again, but they came up against a heroic rearguard action by French and British troops. This was a diversion to give the rest of the Allied troops time to get to the beaches of Dunkirk where a sealift had already begun. The Germans were caught off guard by the evacuation. All they had at their disposal was the Luftwaffe, which harassed the Allied sealift by strafing the soldiers and bombing the ships when the weather permitted. But the RAF flew effective defensive sorties that kept the Luftwaffe from completely dominating the skies over Dunkirk.

By the morning of May 27, the first full day of the evacuation, 28,000 men had been picked up. Many of the stranded soldiers had to wade out into the ocean to reach their rescue craft, a slow and arduous process that also made them vulnerable to the Luftwaffe's sporadic strafing runs. But the weather continued to cooperate. The entire evacuation took place during a freakish, nine-day period of calm, cloudy weather punctuated by fog. As Admiral Ramsay reported, "It must be fully realized that a wind of any strength in the northern sector between the southwest and northeast would have made beach evacuation impossible. At no time did this happen."

In fact, the weather was so calm that the British Admiralty took the unprecedented step of using two stone-and-concrete breakwaters (previously considered too dangerous for docking) as piers to load even more soldiers. It took four hours for the HMS *Sabre* to load 200 troops wading out from the beach; from the improvised piers, larger craft could now load 1,000 troops per hour. And still the low-pressure system that had stalled over Dunkirk delivered calm seas and clouds. The Admiralty upped the ante: perhaps they could attempt the impossible and leave not a single soldier behind.

In England, the navy put out a renewed call for assistance and by May 31, hundreds of private boats — motor boats, lifeboats, barges, pleasure craft and fishing trawlers — made the crossing to Dunkirk. With their help, every last soldier was evacuated by June 4, some 338,226 Allied troops. Because most of these troops had seen action against German soldiers, they became an invaluable military asset to the Allied forces, and Hitler would rue the day they slipped beyond his grasp. On June 5, the wind returned to Dunkirk, but the waves that now broke on shore disturbed only a long empty beach where a few German officers stood, their binoculars trained on the misty horizon over which their enemy had escaped.

RUSSIA REDUX

After capturing western Europe, Hitler turned his sights on Russia. On June 22, 1941, he abrogated the non-aggression pact he had signed with Stalin and invaded with the largest land army the world had seen. The campaign, Operation Barbarossa ("the red barber"), was to be finished by early fall, but the Russians put up fiercer than expected resistance. The German advance under the command of Field Marshal Fedor von Bock didn't reach the outskirts of Moscow until the beginning of December. Hitler, like Napoleon before him, had been expecting a short campaign and had failed to provide his Wehrmacht winter clothing.

Stalled just outside of Moscow, the German troops dug in and predictably began to suffer the frosty wrath of the Russian winter. On November 30, the temperature dropped to -45°C. As Stalin wryly observed, "General Winter has arrived." In December, temperatures regularly fell to -28°C and sometimes as low as -41°C. Inoperative engines grounded the Luftwaffe, aviation fuel lines froze solid and even trucks and tanks were immobilized by the cold. The Wehrmacht reported 130,000 cases of frostbite. The Russians regrouped and counterattacked on December 5, eventually driving the Germans all the way back to Smolensk in early 1942.

But the Germans were obstinate. Operation Barbarossa continued and in the summer, the Germans invaded Russia again, this time from the southern flank. General Friedrich Paulus's 4th Panzer division, on its way to take Stalingrad, had enjoyed tremendous early success in their summer campaign. Unlike the wet August of 1917, the weather had been perfect, and the Panzer division advanced so quickly that they outstripped their supply lines and were forced to stop to wait for fuel. It was frustrating for Paulus because they were ahead of schedule and within striking distance of Stalingrad by the first of September. The Russians, blindsided by the speed of the Panzer advance, didn't have enough time to mount an adequate defense. They knew the city could be lost; if Stalingrad fell, Moscow would most certainly fall as well. But as it happened, the late summer halycon days played into the defenders' hands.

Paulus decided to give his war-weary troops a weeklong furlough to enjoy the sunny, warm weather. They had to wait for supplies anyway, and they had easily bested the Russian troops they'd encountered on the way. Surely they could afford a rest, especially as other German divisions were already engaging the Russians on the outskirts of Stalingrad.

Almost certainly, the furlough cost the Germans Stalingrad. It gave the Russians just enough time to reinforce the city before Paulus joined the battle on September 7. By early October, the Germans controlled 80 percent of Stalingrad, but again the weather came to Russia's rescue. Heavy October rains bogged down the German supply convoys and then, in a second stroke of meteorological fortune, the rain turned to snow on October 19. The winter that followed played into Russian strategies, just as it had the year before on the outskirts of Moscow.

By November, the Soviets had surrounded the German army in Stalingrad and the only way of delivering supplies to the embattled German troops was by airlift. But January was unkind to the Luftwaffe. Temperatures regularly dropped below -30°C and hundreds of German resupply aircraft were disabled by the cold. Although the winter of Stalingrad was not as terrible as Moscow's winter had been the year before, there were a sufficient number of extremely frigid days in December and January that the German soldiers who hadn't been shot, captured or starved simply froze to death in their redoubts. If the summer weather hadn't been so fine and if winter hadn't arrived so early, it is likely the Germans would have gone on to take the rest of the country. Despite the terrible toll of the siege of Stalingrad, the Russians held on for 900 days, and Hitler's occupation plans were ultimately thwarted.

THE SUMMER OF LOVE

Not all the reversals wrought by weather involved military invasions, thwarted or otherwise. Sometimes the weather gods smiled on the cultural affairs of humans. For example, during June, July and August in 1967, a series of high-pressure systems parked themselves over North America and Europe and provided the perfect growing medium for a social revolution that had started months before in the Californian winter.

On January 14, 1967, an extraordinary "gathering of the tribes," as the *San Francisco Oracle* referred to it, took place in Golden Gate Park, San Francisco. It was an unprecedented development in American youth counterculture. The organizers called it the first Human Be-in, and the event brought together the various hippie chapters from all over California and beyond, dressed in their counterculture regalia for a spontaneous celebration of their "freak" identity. The Be-in was partly to protest the criminalization of LSD by the state of California and partly to show the flag. Some 20,000 hippies attended, and the media took notice.

John Phillips of the Mamas & the Papas was there and was so inspired by the heady zeitgeist that he became one of the central architects of the psychedelic revolution. Over the next few months, he wrote the movement's first anthem, a song that Scott McKenzie released as a single on May 13, 1967 — "San Francisco (Be Sure to Wear Some Flowers in Your Hair)." By July, it was at the top of the charts in the U.S. and Canada and went on to become number one in England and much of Europe, eventually selling seven million copies worldwide. Talk about an invitation. Three days after the song was released, the temperature in San Francisco reached 30°C. The normal high for that time of year was 19°C. School ended in late June and the hippies descended on the city. And not just from the United States. It's estimated that 75,000 souls migrated to San Francisco, Berkeley and the Bay area that summer, looking to join a revolution that was poised to sweep the world.

Now that John Phillips had written the anthem, he realized another bigger flagship event was needed to kick off the summer. With Lou Adler, he coproduced a music festival for the tribes in Monterey, south of San Francisco, that ran three days, from June 16 to 18. The Monterey International Pop Festival showcased Jefferson Airplane, the Jimi Hendrix Experience and Ravi Shankar among others. By the last day of the festival, the audience was 60,000 strong and segments of the concert were rebroadcast on television around the world. The Summer of Love, as it became known, had begun.

The timing was extraordinary. Two weeks before the Monterey festival, on June 1, the Beatles released *Sgt. Pepper's Lonely Hearts Club Band*. It turned out to be the psychedelic summer's soundtrack. Also that June, Van Morrison released "Brown Eyed

Girl" featuring a couple making love in a grassy field behind a stadium. Sensuality al fresco.

Outdoor rock festivals and spontaneous outdoor gatherings called love-ins sprang up like psilocybin mushrooms all over North America and Europe, turning the Summer of Love into the greatest open-air party the world had seen. Thousands of aspiring hippies were suddenly thumbs-out on roadsides everywhere, hitchhiking from New York to Vancouver and from Amsterdam to Morocco. Any patch of nature, from a park to a beach to an empty field became the living room and bedroom of these nomadic revelers. The summer landscape was transformed into a vast, outdoor living room: ecstatic, glamorous and lascivious.

And the high-pressure cells delivered perfect weather. In the American Midwest, June was very hot, with an average month-long temperature in Chicago of 26°C. June was also hot on the East Coast, from Halifax to Florida. July wasn't abnormally torrid that year in North America, but then again it didn't have to be. July is always hot.

England and northern Europe had an unusually sunny and warm summer as well. Both June and July were unseasonably sunny, and August was not only drier than usual, it became warmer toward the month's end. In August, the English band Small Faces released "Itchycoo Park," a song about skipping school and getting high in a park. But the best was saved for the last month of the Summer of Love, where it all began, in San Francisco.

Northern California's August and September had been somewhat cooler than normal, but by that time the denizens of the Bay Area were in such an altered state they might not have noticed. The thermostat got cranked up again for October, with an average high temperature of 23°C, in a month that usually has difficulty reaching 21°C. Even November had almost 21 days with a higher-than-average temperature. It's as if the Summer of Love just didn't want to let go.

THE LONG, HOT SUMMER

The series of summer highs that dominated the weather over North America that year had quite different consequences for another demographic of inner-city dwellers. Large metropolitan areas create their own weather if they are big enough, referred to by meteorologists as thermal islands. The heat absorbed by concrete buildings and asphalt surfaces during the day is radiated out at night, raising urban temperatures until they are significantly warmer than adjacent rural areas. This "heat island" effect means that at night, a large city is usually 2.9°C warmer than the surrounding countryside. On calm nights, the temperature differential is often augmented over urban areas by a thermal inversion that traps pollution and heat in a dome. A comfortable night in the country at 19°C would be closer to 22°C in the city. Add in a humidex, which is usually high during a midwestern summer, and downtown you have subjective temperatures of 28°C. That's sweating-in-your-sheets-at-night hot. So the Summer of Love was not quite as pastoral for the citizens of impoverished urban neighborhoods with little access to air conditioning. Their ghettoes became claustrophobic ovens that eventually reached the ignition point in a series of riots that became known as the "long, hot summer." Decades of inequity and oppression had reached the boiling point.

Throughout June, riots broke out in Cincinnati, Buffalo, New York and Tampa. In July, there were even bigger riots in Milwaukee, Minneapolis and Newark. Yet none of them compared to the insurrection that took place from July 23 to 28 in Detroit: the largest urban riot in American history, the 12th Street Riot.

On Saturday night, July 22, Detroit was sweltering, and the downtown residents were outside in numbers to take advantage of the relatively cooler air. At an after-hours club called Blind Pig on 12th Street, a party was in session during the early morning hours of July 23 when a police raid cut the festivities short. They arrested 82 patrons, in some cases using excessive force. Word got out quickly and by the time the police began to ferry their prisoners to the local precinct, an angry crowd of 200 had gathered in front of the club. At some point, an empty beer bottle was thrown through the rear window of a police car, then someone threw a trash can through a store window. Then all hell broke loose.

The violence spread and lasted through the night. The next afternoon was windy. Not good news. Dozens of buildings were burning, and the 25-mile-per-hour winds fanned the flames into raging infernoes. Citizens of Windsor, Ontario, gathered on their side of the Detroit River to watch as great black plumes of smoke began to billow from Detroit's west side. The Detroit police and fire departments were unable to maintain order and the riot spiraled completely out of control.

By Tuesday, July 24, the situation had become international front-page news. Something had to be done. Michigan Governor George Romney ordered in 8,000 National Guardsmen, and a few hours later President Lyndon Johnson mobilized 4,700 paratroopers from the 82nd and 101st Airborne divisions. Even then, it would take another two days to restore order, by which time 43 people had died and 1,400 buildings were burned. In the two years following the riot, 193,000 citizens left the city. Detroit, once the headquarters of a prosperous automotive industry, became an economic disaster zone. The long, hot summer of 1967 had underwritten an entirely different scenario for the citizens of Detroit than it had for those in San Francisco.

POSTSCRIPT

FIRE, WATER, EARTH, AIR

A VISIT TO EARTH'S CORE

Aristotle claimed the world was made of four elements: fire, water, earth and air. It's fair to say his quartet still stands, although it seems that none of them stands wholly alone. The air is filled with water vapor and the oceans hold 50 times more carbon dioxide than the atmosphere does. Deep in the Earth, the rock is on fire, and when volcanoes erupt, they spew finely pulverized lava that drifts high in the atmosphere as ash, sometimes spreading around the world. Earth in the air.

We know there is earth dissolved in the ocean: seawater is 3.5 percent salt and 70 percent of that is sodium chloride (table salt), while the rest contains significant amounts of sulfate, magnesium, calcium and potassium, as well as traces of all the important metals. Indeed, the seas contain a dissolved continent of rock, enough that mining magnesium nodules off the ocean floors could soon be economically feasible. And as far as water in earth is concerned, there are oceans locked far down in the Earth's mantle.

As intermingled as they are, Aristotle's elements are still primarily arranged in a spectrum dictated by mass, with the densest at the center and the lighter elements above. The molten interior of the Earth is heavier than the rocky crust, which in turn is more compressed than water. Water, of course, is denser than air, and air is more substantial than the vacuum of space. Aristotle's elements blend into each other vertically. The earth dissolves into water, the ocean dissolves into the atmosphere and the atmosphere dissolves, ultimately, into space above the thermosphere.

Even though 900°C magma is just 40 miles beneath our feet, it has less thermal impact on us than the sun, which is 93 million miles away. The rocky continents of Earth's surface insulate us from the heat of the interior, so that even though we exist on a relatively thin skin of congealed rock, we are happily unaware of the blast furnace below. Other than the occasional volcano, our hot planetary core usually isn't a main player in the heat economy of the atmosphere. Yet the sticky, flowing magma beneath us contributes to our climate in other more indirect ways.

THE MOHO PROJECT

The seismic waves caused by earthquakes are similar to sound waves — they reflect off any surface they encounter and bounce back. Returning sound waves are what bats use to map their surroundings in the darkness, and by the turn of the nineteenth century, scientists were using simliar principles to analyze the shape and depth of seismic waves from distant earthquakes to map out the interior of the Earth. Perhaps the greatest discovery came in 1909, when the Croatian seismologist Andrija Mohorovičić discovered a discontinuity just below Earth's crust, which is known today as the Moho discontinuity. Using seismic waves, he plumbed the thickness of the continental crust and found that it was about 10 to 60 miles thick, averaging 22 miles thick beneath continents and three to six miles thick beneath seafloors. We are, he discovered, afloat on an ocean of lava.

Of course, this set the stage for the continental drift theory of the mid-twentieth century: that the immovable continents weren't just drifting, they were also colliding. The theorists also claimed some continents were being sucked into the Earth's core in regions they called subduction zones. Despite much initial skepticism, they were proven right. Continents are part of larger pieces of crust called tectonic plates, and these are propelled by convection flows, like slow-motion sea currents, in the viscous magma beneath them.

In the early 1960s, *National Geographic* was pretty exciting. At that time, there were two ambitious races underway: the X-15 project had pilots flying rocket planes to the edge of space; the other, Project Mohole, featured a drilling rig off the coast of Mexico that was chewing through the Earth's crust to reach deep into the magma below. *National Geographic* covered both, and both fascinated me, but Project Mohole especially. I had seen the movie *Journey to the Center of the Earth* and could imagine exactly what it would be like when they reached magma — like an oil well gusher only with lava. How would they cap such a gusher? I envied the workers on the deck of the drilling ship. They might start a new volcano!

The idea behind the Mohole was sound. According to Mohorovičić, the crust would be thinner at the bottom of the ocean. All they had to do was lower a drill from a stationary ship through 14,000 feet of ocean then drill through a mere 17,000 feet of crustal rock. Unfortunately, they didn't have the stabilization equipment that drilling rigs use today. One oceanographer said it was like "trying to drill a hole in the sidewalks of New York from atop the Empire State Building using a strand of spaghetti." By 1966, the project was abandoned after digging a paltry 601 feet into the seafloor.

But the Cold War was at its peak and the Russians took up the challenge. Four years later, in 1970, they started their own deep-drilling project. They might have lost the race to the moon, but they would be the first to get to the center of the Earth. They proved to be much more persistent, and they had the advantage of drilling on dry land, which made their equipment more stable. But after 10 years, they too abandoned their quest for magma. Still, they had bested the Americans. Their drills reached a respectable depth of 7.6 miles, on top of which they discovered two unexpected phenomena: at six miles down, the temperature was already much hotter than anyone anticipated, 180°C; even weirder, the rock was sopping wet.

So water lurks within rock, rock dissolves into water, water evaporates into air, and air, in the form of carbon and nitrogen, gets trapped in rock. And all of it passes through Vulcan's fiery forge deep beneath the Earth's surface.

JOURNEY TO THE CENTER OF THE EARTH

I was nine when I saw *Journey to the Center of the Earth* in 1960. The film follows a group of four intrepid explorers — a famous geologist, his student, a beautiful philanthropist (who funds their exploration) and a tall Icelander with his pet duck — as they wend their way to the center of the Earth following a series of marks left by a previous explorer. Naturally, a rival team led by an evil scientist tries to sabotage them and they barely escape a dinosaur attack, but they eventually arrive at the center of the Earth where they encounter a strong magnetic field and all the metal objects they carry — rings, pocket watches, even their dental fillings — get sucked up into the air.

The movie is based on Jules Verne's novel of the same name, written in 1864. Though richly fanciful, and containing enough science to make it credible to the Victorians, the journey it describes is impossible, given what we now know about the Earth's interior. Direct, manned exploration is completely out of the question, at least for now. But let's take an imaginary ride through the layers of the Earth and see what's down there.

A supernova explosion might seem a strange place to start our journey but consider this: when a big star detonates, one with, say, 20 times the mass of our own sun, most of the energy of the explosion blows off the outer shell of the star. Meanwhile, the inner core implodes in a tremendous gravitational collapse, and all that remains of the original star is a remnant core so dense that electrons and protons are fused together into neutrons. That's not easily done. This dark object, not quite a black hole, can have the mass of two of our suns yet be only seven miles in diameter. It is called a neutron star.

Here's where our thought experiment begins. A teaspoon of neutron star weighs 10 million tons, about 900 times the mass of the Great Pyramid of Giza. If we could enlist Superman to hold that teaspoon and tip it onto the surface of the Earth, it would freefall through the soil and rock and magma like spit through candy floss, reaching the center as fast as a rock falling through clouds. Since the center of the Earth is 3,960 miles beneath our feet, the freefall would take all of 45 minutes and look something like this:

In the first few seconds of freefall, our spoonful of neutron star would cut through the Earth's rocky mantle: the continental crust, the stuff we stand on, what mountains are made of. As Mohorovičić discovered, it's about three to six miles thick under the seabed and about 22 miles thick under the continents. Beneath mountains, it bulges down an extra 15 to 35 miles.

Temperatures rise steadily the deeper you go: 12 to 20 feet beneath the surface of the Earth, the temperature is a constant 11°C. Which explains why our ancestors lived in caves and why a house built into a hill is warmer than one built on flat land. It's hotter still at the bottom of the world's deepest gold mine in South Africa, the TauTona mine. There, 2.4 miles below the surface, the rock face temperature is 60°C. Even with industrial-scale air conditioners going full blast, the ambient temperature for the gold miners is 27°C.

A few seconds later, our tiny dollop of neutron star would hit the upper mantle, just below the rocky crust at about 25 miles down, a depth equivalent to the mesosphere's height above the Earth's surface. This is the lava zone, where temperatures range from 500 to 900°C. Red hot. And it only gets hotter. Together, the upper half of the mantle and rocky crust make up the lithosphere, while the lower half of the upper mantle is called the athenosphere. The lithosphere "floats" on top of the athenosphere. It might take almost a minute for our dense spelunker to go through the upper mantle, which ends about 250 miles below the Earth's surface. After that, it would plunge into the transition zone between the upper and lower mantle that starts at 250 miles below the surface and ends 400 miles beneath that, or roughly the same distance downwards as the thermosphere is above the surface of the Earth. This is the lowest depth that earthquakes have been detected.

Interestingly enough, the transition zone between the lower and upper mantle is thought to contain up to three times as much water as all the world's oceans combined. The magma is saturated like a sponge, a 1,900°C yellow hot sponge to be exact. Oceans in the fiery rock. These thicker layers are taking minutes rather than seconds for our heavy glob of neutron star to traverse, especially the next layer below the transition zone, the lower mantle. It starts at 400 miles and ends 1,800 miles further down.

Below that is the D layer, a name derived from geophysicist Keith Bullen's original term, D double-prime. He coined the tag during the 1960s at the height of geophysics' exploration of the Earth's core. The D layer, extending from 1,800 miles down to 1,900 miles, is thin but active. Here is where thermal fluctuations create hot spots, which propagate plumes of heat that rise through the mantle to the surface.

Finally, at 1,900 miles deep, our stellar spelunker has reached the outer core. The temperature at the boundary of the mantle and the core is around 4,000°C. White hot. The tremendous pressures at this depth, almost three million times those of the surface, paradoxically turn the flowing magma at the bottom of the mantle into solid rock. By contrast, just below that, the outer core is liquid. It is also the source of Earth's magnetism. Like a giant electric dynamo, the metals in the liquid outer core orbit the inner core and generate electromagnetic fields. It is these that give our planet its north and south magnetic poles, and the magnetic field that stretches into space. Solid planets without a liquid outer core, like Mars and our moon, have no magnetic fields. And it turns out that the magnetic field is crucial.

A testament to the fluid nature of the outer core is how many times the Earth's magnetic field has reversed over the past 100 million years — 200 times, or once every 500,000 years. As if we didn't have enough to worry about, it's possible we might be entering a reversal phase right now. Earth's magnetic field has decreased by 15 percent over the past 200 years, and the process appears to be accelerating at a rate of 5 percent per decade. This isn't good.

Earth's magnetic field creates the Van Allen belt, a field of charged particles captured from the solar wind that wraps around the planet just beyond the exosphere like a giant invisible donut with Earth in the middle. Each of the poles pokes out of the donut hole and the inner edge of the hole is clearly visible during auroral displays. The northern lights are the result of charged particles sliding down the edge of the hole in the donut. That's why, from the vantage of the space station, auroral displays are often ring shaped.

The Van Allen belt is more than the source of a beautiful display: it also plays a role in deflecting harmful cosmic rays from pelting down on us, a bit like the ozone layer. If the Van Allen belt magnetic field disappears, the effects on our DNA, on the entire planet's DNA, might be damaging. Alarmingly, there is evidence that Earth's magnetic field is already beginning to collapse in a region in the south Atlantic, between Africa and South America.

Our neutron star explorer has finally arrived at the last layer, the inner core. This sphere at the heart of the Earth is an alloy of solid iron and nickel with its outer edge 3,200 miles beneath our feet. Its center is the Earth's center. Estimates of the temperature here range up to 6,927°C, or equivalent to the surface of the sun. And it has been that way for a long time. Despite the fact that the Earth sheds some 44 trillion joules of heat *per second* via mantle plumes, geophysicists believe that the inner core has cooled only some 400°C during the last four billion years.

Yet it's those mantle plumes that have the most direct relationship to the drifting continents and to the eventual release of naturally sequestered carbon dioxide. Even though carbon dioxide makes up less than 0.04 percent of the atmosphere, it has a disproportionate importance to life. Without it and the other gaseous buffers such as methane and water vapor, Earth's average surface temperature would be -19°C instead of the current 14°C. Obviously, the balance of carbon dioxide in the atmosphere is critical, which is where the magma oceans of the deep mantle come in.

Life processes have changed the composition of the Earth itself, right down to the lower mantle, about 435 miles below us. Carbon capture, or carbon sequestration as it is officially known, is in the news a lot these days. Many fossil-fuel power plants are using a technology that traps carbon dioxide before it is released into the atmosphere. They compress it into a liquid that is then pumped into geological formations deep beneath Earth. Nature has been doing the same thing, on a more massive scale, for billions of years. The world's vast limestone deposits are carbon sequestration on a stupendous scale.

Limestone is calcium carbonate, and it is almost totally composed of the compressed deposits of shallow tropical seas — fossil shells, fossil reefs, fossil algae and fossil oolites (spherical crystals of calcium carbonate that form in warm sea water). Perhaps one of the world's best-known deposits of calcium carbonate are the White Cliffs of Dover. They represent millions of years of stable deposition in the oceans of the Cretaceous era, when tyrannosaurs walked the land and giant mosasaurs plied the oceans.

A six-inch cube of chalk from the Dover cliffs sequesters more than 35 cubic feet of compressed carbon dioxide. And think about all the limestone that has been produced on our planet since the first stromatolites began to deposit calcium carbonates three billion years ago. To get a better idea of the sheer mass of limestone, let's travel to the Bahama Banks, which started forming 150 million years ago, just after Africa detached from North America.

The Bahama accretionary platform is one of the most seismically stable regions on the planet. Dinosaurs came and went while the Bahamas just kept depositing calcium carbonate in shallow tropical seas. Layer upon layer upon layer. So just how deep do these calcium carbonate deposits go? The Deep Sea Drilling Project drilled two cores in 1970. The first bore hole, near Andros Island, went to a depth of 15,600 feet, while the other bore hole, near Cay Sal Bank, reached a depth of 18,906 feet. Neither hit the bottom of the carbonate sediments, though they did get to early Cretaceous era limestone, 140 million years old. That's just the Bahamas. An awful lot of oceanfloor sediments have already been subducted under the continental plates, and with them untold gigatonnes of sequestered carbon dioxide. It runs deeper than you can imagine. And volcanoes, it turns out, occasionally spew out clues to just how deep.

Xenoliths, something that geologists describe as an intrusion or rock trapped inside another rock (usually magma), are carried up to the surface by volcanoes, and by analyzing them, geologists can get a good idea of what is going on in the mantle. Xenoliths come in many forms, but diamonds are the most famous. And some very special diamonds, while worthless for the diamond trade, are invaluable for geologists because they enclose intrusions — almost like xenoliths inside xenoliths — that contain pristine, unmelted samples of the deep mantle, 435 miles beneath the surface. Some contain water, carbon and even ocean sediments.

These relics of ancient seafloors are three billion years old. Yet even more astounding than their age is how deeply they have penetrated the Earth. The deep mantle, once composed of the same rock as the rest of the planets in our solar system, has been altered in chemical composition by life itself. Only the planet's core remains untouched. It has taken some time but life has created our atmosphere and has now reconstituted almost a quarter of the rocky substance of our planet. Earth is being transformed.

COLLIDING CONTINENTS

On a grand timescale, the magma beneath the continents acts like an ocean, with eddies, currents, upwellings and even whirlpools. In a sense, the continents are like clouds: just as the atmosphere both supports and propels clouds, so does the liquid magma interior of the Earth support and propel continents. They drift around on the surface of the magma-like rafts, steered by the currents and upwellings from the fiery depths. All the continents except one have a deeply submerged base that extends into the magma. Like the blocky underwater hull of a barge, the bases act to catch the flow of the magma and carry the continent along.

Sometimes another player interferes with the continents as they bob along on the athenosphere. These are 375- to 500-mile-wide upwellings of magma, called mantle plumes, which rise from the D layer, 1,800 miles down, and climb through the mantle to the lithosphere, sometimes burning through it like a blowtorch. Mantle plumes are relatively stationary. A good example is the Hawaiian Island chain, part of the Pacific plate which is moving west-northwest at three to four inches a year. The Hawaiian mantle plume creates a hotspot that occasionally melts through the crust and erupts. The volcanic eruption forms an island that is then carried away by plate tectonics before the process repeats itself. An even larger mantle plume lurks beneath Yellowstone National Park and is the heat source for Old Faithful.

One of the longest lasting mantle plumes currently sits under the island of Réunion in the South Pacific. A hundred million years ago, it ignited volcanoes that helped to break up the Gondwana supercontinent (which included today's southern landmasses — Antarctica, Australia, South America, Africa and India). It also burned the bottom off the Indian plate after it had detached from Gondwana. Because the Indian plate was half as thick as most continental plates, it skidded over the athenosphere more quickly, covering the 1,900 miles between Gondwana and the Eurasian plate in less than 50 million years. The resulting impact was a tremendous bang-up. The Himalayas and the Tibetan plateau, the highest landmasses in the world, are still rising as a result.

As the two continental plates continue to grind together and buckle in a massive, snail's pace collision, their effect on climate has been equally massive. Indeed, the Himalayas and the high Tibetan plateau affect weather as far away as Australia in a seasonal phenomena called the Indian monsoon, the largest and wettest of the world's monsoons. During the four months from June to late September, India receives 80 percent of its annual total rainfall and oftentimes monumental deluges.

The driver of the whole cycle is the Himalayas mountain range and the high Tibetan plateau. When this region heats up in summer, it draws moist air from the Indian Ocean from as far away as Australia. The inflow is so strong that ocean currents are created beneath the monsoon winds that follow them. These currents are, in effect, underwater winds rather than true ocean currents. Every year, they dredge up plankton from the depths near the Maldives, and for a few weeks hundreds of manta rays congregate in an astonishing and graceful feeding frenzy as a result.

The relationship between ocean currents and wind and climate patterns is quite intimate. The Coriolis effect not only deflects the winds but also ocean currents, causing the great gyres of the north Pacific and Atlantic oceans to rotate clockwise, while those in the southern hemisphere rotate counterclockwise. These gyres are vast currents that shunt warm and cold water thousands of miles, affecting huge swaths of climate. The Atlantic gulf stream supports palm trees on the west coast of Scotland while in the southern Pacific, the Humbolt current delivers cold water to the west coast of South America.

EL NIÑO

It was Ecuadorian fishermen who first alerted meteorologists to El Niño, the warm Pacific current that would sometimes appear around Christmas. Normally the coastal waters of Ecuador are cool, supplied by the nutrient-rich Humbolt current. Plankton populations explode in this cold water, and the anchovies that feed on them thrive. But when El Niño's current brings warm water to the west coast of South America, the anchovy population collapses and affects the fishermen of Ecuador and Peru along with all the larger fish, sea birds and seals the swarming anchovy feed. The effects of El Niño are not confined to the eastern Pacific region. In the last few decades, meteorologists have begun to understand the complex relationship between wind patterns and ocean currents and have realized that the reach of El Niño is long indeed.

Again, it all depends on the trade winds. Normally they blow at a steady rate from east to west across the Pacific, pushing warmer surface waters westward toward Australia and Indonesia. This forces the cold, deep water of the Humbolt current to well up off the coast of South America, as it replaces the eastern wake of displaced warm water. But sometimes the trade winds lessen in intensity, and the region of warmer water is not pushed west at its usual rate, with the result that it backtracks until it reaches South America. This is El Niño. In an El Niño year, as the ripple effect spreads, there may be flooding in Peru, while Brazil, India, Australia and Indonesia experience droughts. In North America, the winter will be warmer in the northeast and wetter in the south.

La Niña is the mirror opposite of its brother: stronger than normal trade winds push the cooler water westward into the equatorial Pacific, cooling surface temperatures and changing weather patterns radically. It brings drought to Peru and flooding to Australia. Together, El Niño and La Niña are referred to as the Southern Oscillation. Weather patterns over half the Earth's surface are influenced by this cycle, and it shows how ocean currents intermingle with the air above to form a supertroposphere, one that extends from the bottom of the deepest ocean trench up to the edge of the stratosphere.

GAIA

As I've researched and written this book, I've come to understand more completely the vast, interconnected systems of oceans, atmosphere and continents and how they are animated by fire — the Earth's heat and the sun's radiation. But I've also been struck by the influence that life has had on these systems. Despite radical changes to the ratio of gases in the atmosphere over the past billion years, the surface temperature of Earth has been maintained within a very narrow range. Exactly the range required by life. There have also been several ice ages during that span, but every time the Earth has recovered, even from Snowball Earth. Furthermore, despite the constant inflow of minerals deposited by rivers, the salinity of the oceans has remained constant for hundreds of millions of years. These self-regulating constants are hard to explain without taking into account the influence of organic processes.

According to James Lovelock, it is life that has tweaked our atmosphere for billions of years, life that has determined the oceans' salinity. In his 1979 book *Gaia: A New Look at Life on Earth*, he set out his theory about how life processes are inextricably intertwined with the non-biological realms of the atmosphere and the oceans. The inspiration for this theory came years earlier, in 1965. He later recalled the exact moment: "An awesome thought came over me. The earth's atmosphere was an extraordinary and unstable mixture of gases, yet I knew that it was constant in composition over quite long periods of time. Could it be that life on earth not only made the atmosphere, but also regulated it — keeping it at a constant composition, and at a level favorable for organisms?" It was a eureka moment of the first degree.

He named his theory the Gaia Hypothesis, after the Greek goddess who personified Earth. Lovelock's theory has been challenged by some scientists who claim that atmospheric levels of nitrogen as well as the Earth's surface temperature and the salinity of the ocean are independent of life processes. The debate is ongoing. Yet I think that whether or not life is regulating or maintaining Earth's surface temperature or the composition of the atmosphere, it is simply enough to acknowledge that life has profoundly transformed all of these processes. Life isn't just the skin of our planet; it has insinuated itself into Earth's geological pulse, not only using it but also shaping it, along with the oceans and atmosphere beyond. This is the unprecedented truth that transcends any quibbles about self-regulating planetary systems. It is to acknowledge the extraordinary power of life itself.

APPENDIX

MEASUREMENT CONVERSIONS

(Some figures have been rounded.)

INTRODUCTION

20 miles per hour = 32 kilometers per hour
29.8 inches = 1009 millibars
50 miles per hour = 80 kilometers per hour

1. STORMY WITH A CHANCE OF LIFE

10 grams = 0.35 ounces
80–90 grams = 2.8–3.2 ounces
2 kilograms = 4.4 pounds
200 kilograms = 441 pounds
53 tons = 48 tonnes
50,000 tons = 45,360 tonnes
1,000 feet = 305 meters
1 square foot = 0.0929 square meters
130–230 megatonnes = 143–254 megatons
332 gigatonnes = 365 gigatons
439 gigatonnes = 484 gigatons
29 gigatonnes = 32 gigatons

771 gigatonnes = 850 gigatons
180 megatonnes = 198 megatons
375,000 trillion gallons = 1,420 trillion liters

2. THE WILD BLUE YONDER

5,200 million million tons = 4,717 million million tonnes
25 million tons = 23 million tonnes
2,000 miles = 3,219 kilometers
1,000 miles = 1,609 kilometers
0.5 miles = 0.8 kilometers
18 miles = 29 kilometers
123° Celsius = 253° Fahrenheit
-181° Celsius = -293.8° Fahrenheit
19 miles = 30.5 kilometers
29,029 feet = 8.851 kilometers
1.5 miles = 2.4 kilometers
35–40,000 feet = 10.6–.12 kilometers
23,018 feet = 7 kilometers
328 feet = 100 meters
90,000 cubic feet = 2,548 cubic meters
1 Celsius degree = 1.8 Fahrenheit degrees
3 miles = 4.8 kilometers
330 miles per hour = 530 kilometers per hour
30 inches = 1015.92 millibars
2 inches = 67.7 millibars
0.5 miles = 0.805 kilometers
11.5 inches = 29 centimeters
26,400 feet = 8 kilometers
-23° Celsius = -10° Fahrenheit
7 miles = 11.2 kilometers
-60° Celsius = -76° Fahrenheit
18 miles = 29 kilometers
13 miles = 21 kilometers
12–19 miles = 20–30 kilometers
0° Celsius = 32° Fahrenheit
14 miles = 22 kilometers
20 miles = 32 kilometers
-3° Celsius = 26.6° Fahrenheit
31 miles = 50 kilometers
50 miles = 80 kilometers
19 miles = 30.5 kilometers
45,000 feet = 12.8 kilometers
120,000 feet = 35.4 kilometers
217,000 feet = 66 kilometers
314,750 feet = 95 kilometers
50 miles = 80 kilometers
160 miles = 240 kilometers
500–2,000° Celsius = 932–3,632° Fahrenheit
200 miles = 322 kilometers
240 miles = 386 kilometers
1 mile = 1.6 kilometers
40 miles = 64 kilometers
3,000° Celsius = 5,432° Fahrenheit
770 miles per hour = 6 kilometers per second
430 miles = 692 kilometers
6,200 miles = 9,978 kilometers

3. CLOUD NINE: INSIDE THE MISTY GIANTS ABOVE OUR HEADS

38° Celsius = 100° Fahrenheit
0° Celsius = 32° Fahrenheit
10 miles = 16 kilometers
-40° Celsius = -40° Fahrenheit
2 Celsius degrees = 3.6° Fahrenheit
1,000 feet = 304 meters
18° Celsius = 64° Fahrenheit
9,000 feet = 2,743 meters
3,280 feet = 1,000 meters
30° Celsius = 86° Fahrenheit
16,400 feet = 5 kilometers
-40° Celsius = -40° Fahrenheit
18,000 feet = 5,486 meters
42,240–52,800 feet = 13–16 kilometers
16,500–50,000 feet = 5–15 kilometers
6,500 feet = 2 kilometers
12–15 miles = 20–25 kilometers
47,000 feet = 14,325 meters
-50° Celsius = -58° Fahrenheit
10,000 feet = 3,048 meters
6,000 feet = 1.8 kilometers

4. THE POEM OF EARTH: RAIN

0.5 millimeters = .019 inches
4.5 miles per hour = 2 meters per second
5 millimeters = .197 inches
20 miles per hour = 9 meters per second
125 miles per hour = 201 kilometers per hour
9 millimeters = .354 inches
50 square miles = 80 square kilometers
18 inches = 45.7 centimeters
20 feet = 6 meters
28 inches = 71 centimeters
44° Celsius = 111° Fahrenheit
1,000 feet = 304.8 meters
350 million tons = 318 million tonnes
12 million tons = 11 million tonnes
15 feet = 4.5 meters
200 miles = 322 kilometers
-40° Celsius = -40° Fahrenheit
350 miles = 600 kilometers
180 pounds = 82 kilograms
80 pounds = 36.2 kilograms
60,000 feet = 18.3 kilometers
50 pounds = 22.6 kilograms
300 hundred square miles = 482.8 square kilometers
250,000 acre-feet = 308,250,000 cubic meters
2,500 miles = 4,023 kilometers
138 inches = 350.5 centimeters
460 inches = 1,168 centimeters
74 inches = 188 centimeters
1,042 inches = 2,646.6 centimeters
-1° Celsius = 30° Fahrenheit
1° Celsius = 33.5° Fahrenheit
4 inches = 10.1 centimeters

5. THE SECRET LIFE OF STORMS

14° Celsius = 57° Fahrenheit
19,000 miles = 30,000 kilometers
37,000 miles = 60,000 kilometers
932 miles per hour = 1,500 kilometers per hour
6,500 feet = 1.2 miles
30,000° Celsius = 54,000° Fahrenheit
124 miles = 200 kilometers
25–56 miles = 40–90 kilometers
31 miles = 50 kilometers
245 miles = 400 kilometers
8 inches = 20 centimeters
6,000 miles = 9,656 kilometers
100 miles per hour = 161 kilometers per hour
2 pounds = 0.9 kilograms
7 inches = 17.7 centimeters
2.5 pounds = 1.1 kilograms
105 miles per hour = 169 kilometers per hour
40–70 miles per hour = 64–113 kilometers per hour
261–319 miles per hour = 420–513 kilometers per hour
150 miles per hour = 241 kilometers per hour
100 miles per hour = 160 kilometers per hour
2 square miles = 6 square kilometers
10,000 feet = 1.9 miles
2–3 miles = 3–5 kilometers
35 miles per hour = 55 kilometers per hour
5 miles per hour = 8 kilometers per hour

6. KATRINA: THE LIFE STORY OF A HURRICANE

25–38 miles per hour = 20–34 kilometers per hour
2,408 miles = 3,875 kilometers
26.5° Celsius = 80° Fahrenheit
150 feet = 50 meters
1,600 miles = 2,600 kilometers
39 miles per hour = 63 kilometers per hour

2.3 miles = 3.7 kilometers
26.05 inches mercury = 882.15 millibars
27.5 inches mercury = 931.25 millibars
230 miles = 370 kilometers
74–95 miles per hour = 119–153 kilometers per hour
155 miles per hour = 249 kilometers per hour
175 miles per hour = 282 kilometers per hour
4–5 feet = 1.2–1.5 meters
18 feet = 5.5 meters
42 feet = 13 meters
155 miles per hour = 249.4 kilometers per hour
40 miles = 64 kilometers
300 miles = 500 kilometers
10,000 feet = 1.9 miles
500 feet = 152 meters
25–28 feet = 7.6–8.5 meters

7. PALACE OF THE WINDS

39 feet = 12 meters
26 feet = 8 meters
10° Celsius = 18° Fahrenheit
3,281 feet = 1,000 meters
40° Celsius = 72° Fahrenheit
12 inches = 30.5 centimeters
-20° Celsius = -4° Fahrenheit
7° Celsius = 44.6° Fahrenheit
-48° Celsius = -54° Fahrenheit
57° Celsius = 103° Fahrenheit
9° Celsius = 49° Fahrenheit
93 miles per hour = 150 kilometers per hour
1.15 miles per hour = 1.85 kilometers per hour
75 miles per hour = 120 kilometers per hour
200 miles per hour = 322 kilometers per hour
100 miles per hour = 160 kilometers per hour
32,000 feet = 10,000 meters
402 miles per hour = 250 kilometers per hour
310 miles per hour = 500 kilometers per hour

8. WHICH WAY THE WIND BLOWS: THE STORY OF WEATHER FORECASTING

34 feet = 10.3 meters
42 feet = 13 meters
32 inches = 80 centimeters
125 square miles = 201 square kilometers
30.9 inches = 1,045 millibars

9. APOLLO'S CHARIOT: THE SEASONS

93 million miles = 149.6 million kilometers
67,000 miles per hour = 108,000 kilometers per hour
0.7 miles per hour = 1.2 kilometers per hour
17 miles per day = 29 kilometers per day
11 miles = 17 kilometers
1 mile = 1.6 kilometer
6,000 miles = 9,656 kilometers
278 million miles = 299 million kilometers
584 million miles = 940 million kilometers
21° Celsius = 70° Fahrenheit
2,000–3,000 feet = 609–914 meters
37° Celsius = 99° Fahrenheit
38º Celsius = 100º Fahrenheit
58º Celsius = 136º Fahrenheit
30° Celsius = 86° Fahrenheit
15° Celsius = 59° Fahrenheit
34° Celsius = 93° Fahrenheit
25° Celsius = 77° Fahrenheit
42° Celsius = 108° Fahrenheit
7° Celsius = 6° Fahrenheit

800 miles = 1,287 kilometers
400 miles = 643.7 kilometers

10. A COLD PLACE: WINTER AND THE ICE AGES

-10° Celsius = 14° Fahrenheit
-15° Celsius = 5° Fahrenheit
-22° Celsius = -7.6° Fahrenheit
35° Celsius = 95° Fahrenheit
18° Celsius = 65° Fahrenheit
-20° Celsius = -4° Fahrenheit
-27° Celsius = -16.6° Fahrenheit
-28° Celsius = -14.4° Fahrenheit
35° Celsius = 95° Fahrenheit
32.2° Celsius = 90° Fahrenheit
31° Celsius = 88° Fahrenheit
30° Celsius = 86° Fahrenheit
29° Celsius = 85° Fahrenheit
-20° Celsius = -4° Fahrenheit
20,000 feet = 6 kilometers
-270.45° Celsius = -455° Fahrenheit
6 feet = 1.8 meters
1.5 miles = 2.4 kilometers
8° Celsius = 46° Fahrenheit
-18° Celsius = -0.3° Fahrenheit
-63° Celsius = -81° Fahrenheit
40 miles per hour = 64 kilometers per hour
200 miles per hour = 322 kilometers per hour
-89.2° Celsius = -128.6° Fahrenheit
-32° Celsius = -26° Fahrenheit
-17.2° Celsius = 1.0° Fahrenheit
- 107° Celsius = -161° Fahrenheit
-35° Celsius = -31° Fahrenheit
65 feet = 20 meters
0.6 miles = 1 kilometers
65 feet = 20 meters
10 billion tons = 9 billion tonnes
-30° Celsius = -22° Fahrenheit
-40° Celsius = -40° Fahrenheit
23° Celsius = 73° Fahrenheit
330 feet = 100 meters
65 feet = 20 meters
16 feet = five meters

11. CLIMATE CHANGE PAST AND PRESENT

8 feet = 2.4 meters
6 feet = 2 meters
330 feet = 100 meters
2 miles = 3.2 kilometers
435 miles = 700 kilometers
330 feet = 100 meters
16° Celsius = 29° Fahrenheit
15° Celsius = 27° Fahrenheit
3 feet = 1 meter
5.6° Celsius = 42° Fahrenheit
10.4° Celsius = 50.8° Fahrenheit
1.2° Celsius = 2° Fahrenheit
0.8° Celsius = 1.5° Fahrenheit
771 gigatonnes = 850 gigatons
29 gigatonnes = 32 gigatons
2° Celsius = 3.6° Fahrenheit
11 inches = 28 centimeters

12. WEATHER THAT CHANGED HISTORY

40 tons = 36 tonnes
120 feet = 37 meters
700 miles = 1,126.5 kilometers
-40° Celsius = -40° Fahrenheit
-45° Celsius = -49° Fahrenheit
-28° Celsius = -20° Fahrenheit
-41° Celsius = -44° Fahrenheit
-30° Celsius = -22° Fahrenheit
30° Celsius = 86° Fahrenheit
19° Celsius = 66.2° Fahrenheit
26° Celsius = 78.8° Fahrenheit
23° Celsius = 73.4° Fahrenheit
21° Celsius = 69.8° Fahrenheit
2.9° Celsius = 5.2° Fahrenheit
19° Celsius = 66.2° Fahrenheit
22° Celsius = 71.6° Fahrenheit
28° Celsius = 82.4° Fahrenheit
25 miles per hour = 40 kilometers per hour

POSTSCRIPT. FIRE, WATER, EARTH, AIR: A VISIT TO EARTH'S CORE

900° Celsius = 1,652° Fahrenheit
40 miles = 64 kilometers
93 million miles = 150 million kilometers
10–60 miles = 20–90 kilometers
22 miles = 35 kilometers
3–6 miles = 5–10 kilometers
14,000 feet = 4,267 meters
17,000 feet = 5,181 meters
601 feet = 183 meters
7.6 miles = 12,262 meters
6 miles = 10,000 meters
180° Celsius = 356° Fahrenheit
7 miles = 11.2 kilometers
10 million tons = 9.1 million tonnes
3,960 miles = 6,371 kilometers
3–6 miles = 5–10 kilometers
22 miles = 35 kilometers
15–35 miles = 24–56 kilometers
12–20 feet = 3.6–6 meters
11° Celsius = 52° Fahrenheit
2.4 miles = 4 kilometers
60° Celsius = 140° Fahrenheit
27° Celsius = 81° Fahrenheit
25 miles = 40 kilometers
500–900° Celsius = 932–1,652° Fahrenheit
250 miles = 400 kilometers
400 miles = 650 kilometers
1,900° Celsius = 3,452° Fahrenheit
400 miles = 650 kilometers
1,800 miles = 2,900 kilometers
1,900 miles = 2,890 kilometers
4,000° Celsius = 7,230° Fahrenheit
3,200 miles = 5,150 kilometers
6,927° Celsius = 12,500° Fahrenheit
400° Celsius = 720° Fahrenheit
-19° Celsius = 0° Fahrenheit
14° Celsius = 57° Fahrenheit
435 miles = 700 kilometers
6 inches = 15 centimeters
35 cubic feet = 1,000 liters
15,600 feet = 4,600 meters
18,906 feet = 5,800 meters
435 miles = 700 kilometers
375–500 miles = 600–800 kilometers
1,800 miles = 2,900 kilometers
3–4 inches = 7–11 centimeters
1,900 miles = 3,000 kilometers

SELECTED BIBLIOGRAPHY

Aeschylus. *Tragicorum Graecorum Fragmenta*, Volume 3. *Danaids*. Translated by Stefan Radt. Gottingen: Vandenhoeck & Ruprecht, 1985.

Alberti, Leon Battista. *De Re Aedificatoria*. Translated by Richard Owen. London, 1755.

Ambrose, Stephen E., and C.L. Sulzberger. *American Heritage New History of World War II*. New York: Viking, 1997.

Aristotle. *Meterologica*. Translated by H.D.P. Lee. London: Heinemann, 1952.

Barnett, Cynthia, *Rain: A Natural and Cultural History*. New York: Broadway Books, 2015.

Barnett, Lincoln. *The World We Live In*. New York: Time Incorporated, 1955.

Boia, Lucian. *The Weather in the Imagination*. London: Reaktion Books, 2005.

Bradbury, Ray. *Green Shadows, White Whale: A Novel of Ray Bradbury's Adventures Making Moby Dick with John Huston in Ireland*. New York: William Morrow, 1992.

Brown, Slater. *World of the Wind*. London: Alvin Redman, 1962.

Brydone, Patrick. *A Tour Through Sicily and Malta, in a Series of Letters to William Beckford, Esq. of Somerly in Suffolk*. London: Strahan and Cadell, 1776.

Bryson, Bill. *A Short History of Nearly Everything*. New York: Doubleday, 2003.

Calder, Nigel. *The Weather Machine*. New York: Penguin Books, 1977.

Campbell, David G. *The Ephemeral Islands: A Natural History of the Bahamas*. London: MacMillan Education Ltd, 1981.

Canfield, Donald E. *Oxygen: A Four Billion Year History*. New Jersey: Princeton University Press, 2014.

Chandler, Raymond. *The Midnight Raymond Chandler*. Boston: Houghton Mifflin, 1971.

Cox, John D. *Storm Watchers: The Turbulent History of Weather Prediction from Franklin's Kite to El Niño*. New York: Wiley, 2002.

Dennis, Jerry. *It's Raining Frogs and Fishes: Four Seasons of Natural Phenomena and Oddities of the Sky*. New York: Harper Perennial, 1993.

Drye, Willie. *Storm of the Century: The Labor Day Hurricane of 1935*. Washington, D.C.: National Geographic Books, 2002.

Durlacher, Chris, dir. "Horizon," *Snowball Earth*. BBC Television documentary, 2001.

Early Greek Philosophy. Translated by Jonathan Barnes. London: Penguin, 1987.

Elert, Emily, and Michael D. Lemonick. *Global Weirdness: Severe Storms, Deadly Heat Waves, Relentless Drought, Rising Seas and the Weather of the Future*. New York: Pantheon, 2012.

"The Evacuation from Dunkirk." newworldencyclopedia.org

Fagan, Brian M. *Floods, Famines and Emperors: El Niño and the Fate of Civilizations*. New York: Basic Books, 2000.

———. *The Great Warming: Climate Change and the Rise and Fall of Civilizations*. London: Bloomsbury, 2009.

———. *The Little Ice Age: How Climate Made History 1300–1850*. New York: Basic Books, 2001.

———. *The Long Summer: How Climate Changed Civilization*. New York: Basic Books, 2004.

Fujita, T. Theodore. *Workbook of Tornadoes and High Winds for Engineering Applications*. SMRP Research Paper. Chicago: Dept. of the Geophysical Sciences, University of Chicago, 1978.

Gleick, James. *Chaos: Making a New Science*. London: Penguin, 1987.

Gribbin, John. *Weather Force: Climate and Its Impact on Our World*. New York: G.P. Putnam's Sons, 1979.

Gribbin, John and Mary. *Watching the Weather*. London: Constable, 1996.

Hamblyn, Richard. *The Invention of Clouds: How an Amateur Meteorologist Forged the Language of the Skies*. New York: Farar, Straus and Giroux, 2001.

"The Ice Man Cometh," *York University Magazine*, Fall 2016.

Jankovic, Vladimir. *Reading the Skies: A Cultural History of English Weather, 1650–1820*. Chicago: University of Chicago Press, 2001.

Kals, W.S., *The Riddle of the Winds*. New York: Doubleday & Company, 1977.

Kenda, Barbara, Editor. *Aeolian Winds and the Spirit in Renaissance Architecture: Academia Eolia Revisited*. Abingdon, U.K.: Routledge, 2006.

Ketjen, Jim. "Evidence the Climate May Go Crazy," *Toronto Star*, October 2, 1994.

Klinenberg, Eric. *Heat Wave: A Social Autopsy of Disaster in Chicago (Illinois)*. Chicago: University of Chicago Press, 2003.

Kohn, Edward P. *Hot Time in the Old Town: The Catastrophic Heat Wave that Devastated Gilded Age New York*. New York: Basic Books, 2010.

Kolbert, Elizabeth, "Ice Memory," in Annals of Science, *The New Yorker*, January 7, 2002.

Larson, Erik. *Isaac's Storm: A Man, a Time, and the Deadliest Hurricane in History*. New York: Vintage, 2000.

Levine, Mark. *F5: Devastation Survival and the Most Violent Tornado Outbreak of the Twentieth Century*. New York: Hyperion, 2007.

Lucretius. *On the Nature of the Universe*. Translated by R.E. Latham, revised by John Godwin. Harmondsworth: Penguin, 1994.

Ludlum, David M. *The Audubon Society Field Guide to North American Weather*. New York: Alfred A. Knopf, 1991.

——. *The Weather Factor*. Boston: Houghton Mifflin, 1984.

Lee, Laura. *Blame It on the Rain: How the Weather Has Changed History*. New York: Avon, 2006.

Le Grand, Anthony. *An Entire Body of Philosophy, According to the Principles of the Famous Renate Des Cartes, In Three Books*. London: Samuel Roycroft, 1694.

Lynch, Peter. "The Origins of Computer Weather Prediction and Climate Modeling," *Journal of Computational Physics* 227, 3431-3444, 2008.

Mergen, Bernard. *Weather Matters: An American Cultural History Since 1900*. Lawrence: University Press of Kansas, 2008.

Moore, Peter. *The Weather Experiment: The Pioneers Who Sought to See the Future*. New York: Vintage, 2015.

Mykle, Robert. *Killer 'Cane: The Deadly Hurricane of 1928*. Boulder, C.O.: Taylor Trade Publishing, 2006.

Procipious. *History of the Wars, Books III and IV: The Vandalic*

War. Translated by H.B. Dewing. Cambridge: Loeb Classical Library, Harvard University Press, 1924.

Rankin, William H., *The Man Who Rode the Thunder*, New Jersey: Prentice-Hall, 1960.

Resnik, Abraham. *Due to the Weather: Ways the Elements Affect Our Lives*. Westport, C.T.: Greenwood Press, 2000.

Seneca. *Naturales Quaestiones*. 2 volumes. Translated by T.H. Corcoran. London: Heinemann, 1971.

Shukla, Jagadish, quoted from David Weinberger's "The Machine That Would Predict the Future," *Scientific American*, December, 2011.

Simons, Paul. *Weird Weather*. Boston: Little Brown, 1997.

Singh, Devendraa, A.K. Singh, R.P. Patel, Rajesh Singh, R.P. Singh, B. Veenadhari, and M. Mukherjee. *Thunderstorms, Lightning, Sprites and Magnetospheric Whistler-mode Radiowaves*. http://arxiv.org/pdf/0906.0429.pdf.

St. Clair, Chris. *Canada's Weather: The Climate that Shapes a Nation*. Toronto: Firefly Books, 2012.

Strauss, Sarah. *Weather, Climate, Culture*. Toronto: Perigee Trade, 2010.

Theophrastus of Eresus. *Concerning Weather Signs*, Enquiry into Plants, Volume II. Cambridge: Loeb Classical Library, Harvard University Press, 1926.

Theophrastus of Eresus. *On Winds and or Weather Signs*, Enquiry into Plants, Volume II. Cambridge: Loeb Classical Library, Harvard University Press, 1926.

Thorphe, Edgar, and Showick Thorpe. *CSAT Manual 2012*. India: Pearson, 2012.

Timmer, Reed. *Into the Storm: Violent Tornadoes, Killer Hurricanes and Death-Defying Adventures in Extreme Weather*. New York: Dutton, 2010.

Vitruvius. *The Ten Books on Architecture*. Translated by Morris Hicky Morgan. Cambridge: Harvard University Press, 1914.

Watson, Lyall. *Heaven's Breath: A Natural History of the Wind.* New York: HarperCollins, 1984.

Weather, Firefly Books, calendar, ISBN 978-1-55297-399

"What's Up With the Weather?" *Frontline*. NOVA television documentary, April 18, 2000.

Winters, Harold A. *Battling the Elements: Weather and Terrain in the Conduct of War*. Baltimore: John Hopkins University Press, 1998.

Yeager, Paul. *Weather Whys: Facts, Myths and Oddities*. Toronto: Perigee Trade, 2010.

ACKNOWLEDGMENTS

Firstly, I thank Jack David, Michael Holmes and ECW Press for standing by me all these years. Their continuing support made this book possible. In addition, I thank Susan Renouf, my editor, whose encouraging and thoughtful responses helped guide me through the final stages of this manuscript. I'm thankful also to Rachel Ironstone, who was my managing editor at ECW, along with Susannah Ames, my ebullient publicist and Amy Smith, who handled marketing. I thank my copy editor Crissy Calhoun, who went well beyond the call of duty, as well as Jen Albert, who proofread the manuscript. I am, of course, indebted to Barbara Gowdy for her marvelous ear and sharp eye.

Thanks is due to Bruce Westwood of Westwood Creative Artists who has championed my work for over a decade. Meg Wheeler, also of Westwood, has quietly worked behind the scenes to disseminate my manuscripts. Graeme Gibson responded to excerpts from earlier versions of *18 Miles*, as did Margaret Atwood. Their support was invaluable. Thanks is due also to Chris Scott, chief meteorologist of The Weather Network, who reviewed this manuscript for scientific veracity.

John Donlan of Brick Books provided logistical support during the writing of this manuscript, as did *Brick, A Literary Journal* through the generous agency of the Ontario Arts Council. I must also acknowledge Wikipedia, an extraordinary resource for research, often equaling and sometimes surpassing primary sources. Grateful acknowledgment is due to the authors whose work is cited in the book and listed in the bibliography.

INDEX

D

N

O

X

Z